The Busy, the Bossy & the Bully

incorporating
The NHS, the Uniform
& the Blackberry®

By
Lucy la Zouche

© Lucy la Zouche 2017

Lucy la Zouche has asserted the right to be identified as the author of this work in accordance with the Copyright, Designs and Patents Act, 1988 (as amended)

Web Site: www.lucylazouche.com

e-mail: lucy@lucylazouche.com

 **@LucylaZouche**

ISBN 978-0-244-30105-7

CONTENTS

CONTENTS (Continued)   Page

FOREWORD

Arthur Seldon, the economist, said in his book 'Capitalism' that it is 'the Government of the Busy by the Bossy for the Bully'. My experience over the past few years proves that this applies equally to many parts of our National Health Service (NHS).

First and foremost, this book is about a bullying culture that has become common-place within our NHS. Bullying is insidious and destructive, it undermines the way employees work and it ultimately undermines their happiness and their mental wellbeing.

The Advisory, Conciliation & Arbitration Service (ACAS) says Bullying is: "offensive, intimidating, malicious or insulting behaviour, an abuse or misuse of power through means intended to undermine, humiliate, denigrate or injure the recipient". It goes on to say that it "may be by an individual against an individual or involve groups of people. Whatever form it takes, it is unwarranted and unwelcome to the individual".

This book is faction, fiction based on facts, as was my original book "The NHS, the Uniform and the Blackberry®". There is no intention to put anyone in the stocks so that they can have bad fruit thrown at them. Indeed, you will note that there is only one named person in this book, James a former mental health nurse and trainer who suffered bullying and harassment in the NHS workplace, under the malign influence of his Senior Manager.

This two part book enables newcomers to read James' complete story and it enables the people who read the original book to follow James' rehabilitation and his successful move into a new career. It also highlights to everyone the appalling weaknesses of the registration and regulatory system for nurses and midwives run by the Nursing and Midwifery Council (NMC).

Our great NHS has been revered the world over and it has many frontline staff that care deeply about their patients and regularly work the odd miracle. When we say frontline we, more often than not, think Doctors or Nurses but of course there are many others: Physiotherapists, Psychiatrists, Psychotherapists, Occupational Health, Midwives, Receptionists and cleaners to name a few. They all make the NHS work.

It is not just more and more money that the NHS requires. It needs good, well-trained managers; and, there are far too many layers of management, too many bad examples of empire-building. Just because someone has been a nurse does not mean they are a good manager. Just as each nurse goes through extensive training so should every manager.

The NHS is a business and it requires good leadership; all too often this is lacking. A CEO does not need to know how to perform a surgical operation or bandage a leg although there is no harm if they do. No, they need to know how to manage and motivate the people who do perform these tasks, on a daily basis.

The truth is that, in any business, outstanding specialists do not automatically make good managers. And, if you pay the specialists you desperately need too little and the managers too much, the specialist will only have one way to go.

Sadly, many bullies have found their way into NHS management and some of our NHS Trusts are slowly being destroyed by them. These martinets, the kind of people that take great joy in issuing peremptory orders and disciplining employees rather than coaching and finding reasons to give praise, have no place within our NHS.

Our NHS is a wonderful institution and we should care for it. Just as you might amputate a limb that has gangrene so we should excise these martinets and let the NHS, once again, become the caring organisation it should be.

To be fair, the NHS Trust James worked for is undergoing a complete 'refit'. Directors and senior executives have departed, significant parts of the Trust have been moved into other, better managed

Trusts and a new regime is trying to re-motivate the staff and bring certainty and confidence to the patients and their families.

I would like to dedicate this book to the caring people within the NHS. But, I also want to dedicate it to James who has struggled over the past nearly 5 years to gain control of his life. He is once again happy, stable and contributing to society.

So, this book is based on real-life events which happened to a real person who discovered what it is like to suffer mental health problems when working for the NHS and then dealing with the profession's regulatory body, the NMC.

You would expect the NHS to be switched on to recognising employees suffering stress in the workplace and experiencing the onset of depressive illness would you not. Well think again!

You would expect a nursing regulatory body like the NMC to be at the leading edge of handling people who have suffered pain and trauma in their life. Well dream on!

Reactive depression is a depressed mood state related to stressful life events. James, the subject of this book, suffered more than his share of those stressful life events. Over a period of two and a half years he was bullied, harassed, subjected to an inordinate number of investigations and put through two complex Disciplinary Hearings.

During that time, his beloved grandmother died from a debilitating and extremely painful illness. And, his wife had what could have been a life changing accident. The NHS Trust in question has an impressive written policy dealing with the management of stress and workplace wellbeing. This document clearly sets out the symptoms to look for and the actions to be taken.

James exhibited all the classic symptoms of Workplace Stress and Reactive Depression. His managers just interpreted them as delinquency and bad behaviour and treated him accordingly. There was no compassion, no thoughtfulness, no understanding of his vulnerable condition.

He was finally driven to attempt suicide in a lay-by in his car! Sadly, James and his family have never been able to thank the 'Good Samaritan' who called the police.

It is James and his family that have asked me to recount his story; it was as much a catharsis for them as it is an exposé of the way that staff relations are sometimes handled in the NHS. The conversations I had with the family come at the end of the first half of this book

If anyone in the NHS reads this book and recognises the poor management behaviour, the obsessive pursuit of a vulnerable colleague and the total lack of compassion as something they may have been guilty of, at some time, somewhere, I hope they feel ashamed.

Both the NHS and the NMC seem to have a warped view of effective staff management and a distorted view of openness and fairness. They also have a lingering, Victorian approach to mental illness.

Under the influence of aggressive lawyers, NHS Trusts use the threat of court or tribunal costs to try to avoid their responsibilities to staff and to patients and their families. Similarly, dominated by its vast team of in-house lawyers and paralegals, the NMC acts as an enforcement arm for the NHS Trusts and, primarily, it tends to focus on meting out punishment to the weak and lowly paid whilst avoiding applying any form of justice to people in senior positions earning inflated salaries.

And, unlike James' Senior Manager more and more NHS Trust senior managers are not NMC or GMC registrants; thus, the NMC becomes less and less relevant to patient safety. Who is going to act as regulator to this increasing large body of senior managers who have no clinical qualifications? They are the people establishing policy, making the resource decisions and controlling day-to-day operations but they cannot be sanctioned for their failures by a relevant professional regulatory body.

Lucy la Zouche   2017

Book 1

The NHS, the Uniform & the Blackberry®

Introduction

My name is James and, at the age of 39, I am starting my life again. There are pluses and minuses to that but at least I am alive and cared for; and, once I have completed my rehabilitation, I can contemplate a new career.

As a child, I had a fairly normal but happy life growing up with my Mum and Dad and my elder brother. Sadly, though, my father died when I was approaching 14, a difficult age for a teenager to lose a parent. As a keen skate-boarder and a bit of a lad, the streets became my escape and my boarding friends filled the gap left by my father's death.

I left school with a clutch of GCSEs, but had no idea what I wanted to do; I held down a number of different jobs until I started work as a care assistant in an NHS Mental Health Unit. The work suited me well because my being a bit of a lad went down well with the patients; but, it was just a job not a vocation.

In 1999, I married my teenage sweetheart who had supported me when my father died. As teenagers do, we had drifted apart and then, happily, we drifted together again. She was working to support herself through her doctoral studies and she persuaded me to qualify as a

Mental Health Nurse.

With a steady home life, I qualified in 2003. It was great to receive my nursing diploma the following summer within a couple of weeks of my wife receiving her PhD. We were happy in our marriage and we both settled into our new careers.

I spent three years as a Staff Nurse in Mental Health Units and worked extended hours on agency shifts. The EU Working Hours Directive does specify limits to the hours an employee can be expected to work; but, the NHS could not survive if staff were not prepared to sign waivers and work longer. For me, the extra work was enjoyable and I also felt more worthwhile if I could match my wife's earning capacity.

After three years of full time nursing, I spotted an opening as a Mental Health Nursing Trainer, applied, got the job and became a regular round peg in a round hole.

I liked teaching people and they told me I was good at it; I could continue agency nursing shifts to keep my knowledge current and supplement my income; and, our team was a good place to work.

Over a period of 4 years there were regular management and structural changes but nothing prepared me for the upheaval that took place in 2010.

A merger of NHS Trusts into a new giant

resulted in the arrival of a new boss with a forces background. The whole team was put on notice about their jobs and it was made clear that we were likely to lose one person.

Call me naïve if you like but, although I was worried, I did not anticipate the lengths to which work colleagues would go to make sure they did not lose their jobs; the law of the jungle was alive and well in the good old NHS.

A combination of events followed which destroyed the morale and cohesion of the team, sent my career spiraling down to an eventual dismissal, ruined my mental health and nearly ended my life.

I have suppressed the names of the people and of the Trust involved because, although I hold many of them responsible for my problems, with 20:20 hindsight I also blame myself for repeatedly shooting myself in the foot.

But mainly I blame the NHS.

It is the NHS that employs people as managers who may have relevant professional qualifications but who cannot manage people.

It is the NHS that specialises in writing procedural manuals and guides for everything you can think of but does not ensure that managers read them, understand them or adhere to them.

It is the NHS that thinks big is good and that

mergers will achieve everything when all they do, if badly managed, is to cause confusion, consternation and demoralisation.

It is the NHS that has created a culture where suspension and formal disciplinary action are preferred to coaching, encouraging and motivating.

As a result, no one in our NHS Trust was capable of recognising the symptoms of Workplace Stress and Reactive Depression which were dragging me down.

Several of my colleagues were qualified mental health clinicians and most of the rest had nursing qualifications. Not one of them identified my symptoms or provided the support I really needed. My wife summed my situation up when she said "Where do you hide a tree? - in the forest!"

The title of this book may seem strange to readers; but, when my Senior Manager and my immediate line manager were responding to my criticism that I felt that I had had no support, they both responded that I had been supplied with a new uniform and more up-to-date technology.

No mention was made of what was really missing – humanity, compassion and interest in the individual. It was as though they were saying "What are you complaining about? You got a new Uniform and a Blackberry®!"

New Direction

No sooner had our new boss arrived than we realised that things were going to change. It would have been hard to miss because we received a flood of e-mails telling us how that change would affect us.

It is perhaps a good job that we did not see all the e-mails at that time because we were referred to in pejorative terms such as 'my inherited team'.

Those of us who managed to get on the wrong side of the new boss were also disciplined by lengthy e-mails which were actually in direct contravention of the Trust's Internet and E-mail policy; but, for me, they appeared as a bright red flag to my inner bull!

Another innovation was 'one-to-one meetings' with my boss, more often than not, involving at least one and often two other managers. Everything was formal, notes were taken and a summary e-mail came out with the conclusions, findings or warnings.

Of three active male trainers, one was promoted to provide a three tier 'one-on-one-on-one' management structure for us to navigate and two of us were told that it was possible that, if we did not toe the line, we would be 'performance managed' out of our jobs.

Morale and team cohesion flew out the window. Everyone was looking to save their own necks. Later, my boss described all this as 'firm but fair' management.

My training colleague and I were both told off for the language we used in restraint training sessions even though the idea was to replicate a real life Mental Health environment. We were also accused of being cheeky about our managers to our trainees; well, yes we were, but previous managers had taken this with a pinch of salt.

Where I shot myself in the foot was that my colleague listened, changed his behaviour and became an informer; I took it all personally and started to suffer from stress.

My new boss did not want me breaching the Working Time Directive and, even worse, the waiver form I had signed mysteriously disappeared.

I was ordered to rein back my shift work to one a week and this strictly in accordance with the directive which also sets a time gap between periods of work. I took the instruction on board but had a few commitments in the diary and I did not want to let anyone down.

To add to my problems and increase the pressure upon me, at about that time my wife had a serious accident on her horse and very nearly broke her back; the horse later had to be destroyed because it had a trapped nerve in the neck. My wife was in intense pain for months and I was, effectively, her carer.

In the absence of a competent NHS Physiotherapist, she embarked on private treatment involving physiotherapy, chiropractic treatment and hydrotherapy at considerable expense.

And, just to complete the picture, I should mention that my grandmother, who had done so much to support the family when my Dad died, was herself dying from a particularly unpleasant cancer.

My immediate line manager was aware of what was happening in my life but he clearly did not know what to do about it; from the total lack of interest from above, I can only assume he kept it to himself. At no time, did anyone enquire about my wife, my grandmother – or for that matter the horse. As for me........

At work, I had become the sole focus of my Senior Manager's disciplinary activity. And, the first manifestation of that was that I was admonished for meeting those work commitments which were already in the diary. On reflection, I should have obtained specific permission to do that but, I did not.

I found out later that the investigation could not

prove that I had worked too many hours but only that I had not allowed enough time between work periods..........

STRIKE ONE!

Fraudster!!

My training job involved quite a bit of travelling and so over a six-year period there were a lot of expenses claims. I may never know why my expenses claims for the entire six-year period were examined – it may have been because of something that happened later – but they were.

And, there was one mistake; I had claimed for a journey for a course that was in my diary but had later been cancelled. My immediate line manager had checked the claim and signed the same 'penalty of death' statement of correctness but I was told I was the only one in the wrong. And, the claim was for about £5 which I immediately offered to repay..........

STRIKE TWO.

Around the same time, I came under criticism for finishing courses early and going home and for not turning up on time for courses. I was threatened with a clocking-in process exclusive to me.

Well, in fact, the whole team would go home if a course finished early; and, at my second Disciplinary Hearing, my boss confirmed publicly that going home early in these circumstances was OK.

The business of not turning up on time had

everything to do with my wife's problems and the very few occurrences were cleared with my training 'buddy'.

Two members of the management team were asked to peer-review my training and the results were top class. But what I did not know was that more mini-investigations were taking place.

What I did know was that, rightly or wrongly, I felt bullied and badly stressed. I submitted a grievance against my boss but it was peremptorily batted back without investigation; the union representative who suggested I pursue the grievance had been worse than useless.

Meanwhile, my training 'buddy' had received a photograph from a nursing friend purporting to show me sleeping whilst on night duty in a ward. He showed it to our Senior Manager who initiated a full-scale investigation.

In fact, nurses are entitled to short break on night duty, believe it or not 47.5 minutes, and if they want to they can have a nap. That investigation bit the dust but, in terms of manpower, it was at some considerable cost.

And, someone suggested that, as part of my uniform arrangements, I had claimed for three sets of trainers. That turned out to be just one and the claim was perfectly legitimate.

During that summer, with my mind in a mess anyway, I had a 'close encounter' with an articulated lorry at a roundabout about 20 miles from home. I was on my way to the third day of a training course I had been sent on and, suddenly, I felt that the world had crashed around my ears. I headed home and spent the day pulling myself together.

Shortly afterwards I was hauled on the carpet for taking the day off but, much worse, not reporting the accident to the car leasing company and for claiming the cost of the journey.

I explained that the only damage was to my nerves and that I had actually done the journey and that was why I had claimed it. But it was..........

STRIKE THREE.

The resultant, lengthy disciplinary e-mail from my Senior Manager combined with the recent fuss about timekeeping then led me to a fatal error. My wife had to attend an MRI for her back injury and I knew she could not and would not attend without me. I believed there would be no sympathy and a request for leave would be refused.

Yes, I hear what you say. Someone else in the family or a friend could have attended with her, but I felt it was my duty and I knew she would not attend on her own.

So, I said the appointment was for me and when asked for the appointment letter, I altered the letter. I was advised by my lawyer later that if I had 'thrown a sicky' it would have been fine but what I had done was actually fraud and, therefore, could be gross misconduct..........

OUT????

The fraud team was called in in July and I was put under formal notice of an investigation. I did not feel that I could tell my wife because she was so ill or my family because I was so ashamed. I was like a rabbit in the headlights.

There was much talk about reducing the size of the team to save cost and I was sure that I would be the saving!

The investigation dragged on but, much to my surprise, in October 2011, I was confirmed in my job after a job matching process and all seemed a little bit better in my world.

But, it was a false dawn. The following spring, some nine months after the formal investigation started, I was hauled before a Disciplinary Panel on the charges I have outlined.

The Trust procedures did say that disciplinary investigations should be over in four weeks but the panel said that the delay was 'acceptable' without giving any coherent reason.

I cannot remember the Disciplinary Hearing because I was in a state of extreme stress and shock. Somehow though, and perhaps it was the union representative that saw me through, I ended up with a Final Written Warning rather than a dismissal.

And, I realise now that they could have dismissed me on the forged appointment letter alone. But, my illness had still not been recognised despite the guidance available in the Trust procedures.

To be fair, after reviewing all the paperwork later, the first Disciplinary was properly documented, properly conducted and allowances were made for my foolishness and personal circumstances.

My Senior Manager had followed best practice in keeping at a distance from these investigations but that mistake was not going to be repeated! I could sense a profound unhappiness with the outcome of the Disciplinary Hearing and started to worry about what would happen next.

A Second Bite at the Cherry

And, I was right! Barely had I been able to draw breath but I was under investigation again just seven days later.

Prior to going to France on holiday I had accidentally locked my Blackberry® phone and invalidated the password. So, I phoned my 'training buddy' to get the telephone number for IT. They told me how to reset the password.

My 'buddy', in the best tradition of telling tales out of school and resolutely observing the law of the jungle, suggested to his manager that I must have phoned from France to do this. And, he warned that I was probably going to use my Blackberry® and perhaps my laptop from abroad.

In fact, I left my work Blackberry® and laptop at home and no one could ever prove any misuse of the laptop but it was enough to start a new witch hunt.

When our Blackberry® phones were issued, we should have signed a form to say that we had read the Conditions of Use which prohibited personal use.

The forms were, in fact, sent to everyone attached to a very obscure e-mail and the attachment was entitled 'Application Form'.

Of course, I probably should have opened the attachment, read the form and signed and returned it. But, I did not and an e-mail revealed to me recently and sent after my dismissal shows that most of my colleagues were still being chased, two years later, for a signed form!

The person who handed me my Blackberry® told me that the tariff was all inclusive and I took that at face value. Over a period of about eighteen months, the Trust incurred excess charges of about £20 in total because of my personal use of the phone.

There was no system in place to identify personal use, no early warning to me or to management and it could have gone on for ever had the investigation not been triggered. I wonder how much personal use remains undetected on other Trust mobile phones?

Excuses, excuses, I hear you say but, as I understand it, most large companies monitor this kind of thing closely and their employees get a chance to pay the excess and, at least the first time, receive a reminder of the policy.

If I had been warned when my first excess of a few pence had occurred, that would have been the end of it. If I had carried on there would still have been plenty of time to include the misdemeanour in the first Disciplinary. As it was..........

STRIKE ONE – AGAIN.

The next allegation was bizarre in the extreme.

A young woman, who was on the verge of disciplinary action herself, had been withdrawn from active training duties and had been told by her manager to spend the day sorting and watching training videos. I was working in the same area and had left a feature film DVD on my desk.

That day I had spoken to IT about enabling DVDs to be played on a training room laptop. Whilst I returned the training laptop to the training room, the young woman in question borrowed my DVD and started to watch it instead of the training DVDs.

I was accused of enabling the playing of DVDs on her laptop when, in fact, that was already enabled; they accused me of not supervising her even though she was not my responsibility.

I was supposed to know that her manager, who had a separate office in the building, had left for home because he was not well. There was no handover, my 'training buddy' was her normal supervisor and the manager was responsible for setting her work 'goals'.

The investigating officer interviewed her and, of course, she protected herself by blaming everything on me. But, she would not sign her statement and refused to appear at any hearing. But, guess what..........

STRIKE TWO – AGAIN.

E-mails available to me later showed that the perceived view was that there was not enough substance in these two allegations to justify another Disciplinary Hearing.

The middle manager in my managerial sandwich had given me an e-mail 'slap on the wrist' for the DVD incident and the lion's share of the phone use had been before the first disciplinary. My Senior Manager had a dilemma.

You may think that I am over-egging this but during all of this I was even investigated because a course I was attending to achieve a training qualification had overrun.

For technical reasons the course had to be extended for a couple of weeks but my Senior Manager thought I was lying; my mileage claims were double checked; and, a check was made that I had actually attended all the course sessions.

Incidentally, I got top marks for this course; but, neither my line manager nor my Senior Manager gave this any recognition or offered any congratulations. It was one more opportunity for them to hold an investigation.

The Cavalry Arrive

Just over a month after my first disciplinary a new member joined the team. She came from a social worker background and had to undergo a period of training and acclimatisation before becoming a full-time trainer.

One of my particular areas of expertise was a course on Care and Restraint which was half theory in a classroom and half practical in a gym.

The newcomer had not worked with me in her first month with the team but she was due to join me for a one week course at the end of June. As the experienced instructor, I would demonstrate the holds and she would be my assistant/model.

Come the day, I had already received a message from her that she would be late because of a dental appointment. It emerged in evidence later that that day was also 'her time of the month'; she arrived just in time for the physical part of the morning which was to be conducted in the gym.

She said nothing at the time but it turned out that, the next day, she spoke to my line manager; she accused me of putting her in a fireman's lift in front of the class and of various other 'behaviours' which were not specified as to nature or time and date.

It seems that my very inexperienced junior line manager soothed her with platitudes but, from his evidence given later, he completely failed to do anything about her complaint.

He clearly did not consult the Trust's Harassment and Bullying procedures because he did not speak to me, did not try to resolve the issue via informal mediation and did not enlist any help from his superiors.

But, it took a month for a formal complaint to be filed and at no time was I spoken to. In that time, we had a Team Away day and, somehow, I ended up sitting next door to her.

As I later found out that she said in evidence that it had been "necessary to change her behaviour by trying to keep her distance", and that she "dreaded being near" me, I do ask myself why she did not try to relocate herself by switching with someone else.

Her friend sitting beside her became the sole witness who said that I touched her on the knee; but, as we were at a round table, that must have been quite difficult to see.

This time my line manager was confronted by a complainant and a witness. He had to take some sort of action and, characteristically, passed the buck to my Senior Manager.

Later an investigation commenced and I was put on restricted practice so that I could not be in the same location as the complainant.

By then, we had been in the same workplace for a maximum of ten days and we had actually worked together for six of those days. I was accused of persistent touching and prodding and of the original fireman's lift.

Five delegates on the training course were interviewed and not one of them saw the lift. But, my line manager and 'training buddy' gave almost identical evidence that 'sometime in the past' they had seen me lift a female delegate attending a training course.

Well, I knew a lot of people by then and I might have given a friend a hug but then no-one was saying that it was a fireman's lift. Even my line manager admitted that it gave him no cause for concern at the time.

And, having worked for many years in an environment where the female/male ratio was 80:20, it is perhaps surprising that no one had complained about my behaviour previously.

In the end the only substantive evidence to corroborate hers was that I had touched her on the knee once at the Team Away-day; to be honest I don't know whether I did but I said that it was possible that I had.

Hands up anyone who can say that they have never tried to gain someone's attention by touching their leg or their arm.

Despite Trust procedural requirements, I was

never spoken to informally even by my new female colleague; and, there was never once an attempt at mediation despite the Trust's specific policy requirements. The allegation was taken straight to disciplinary hearing.

My Senior Manager originated an extra allegation which was to show that my training performance had not improved and that I was generally a worthless employee.

No evidence came forward except some extraordinary statistical analysis comparing my performance to that of my training 'buddy'. The samples were skewed and the analysis totally flawed; this was later admitted in a letter written to my lawyer by a senior manager in HR; this last allegation really fell on its face. But, if you throw mud some of it sticks.

I need to remind readers that I had kept my work problems secret from my wife and family during the first set of allegations and I have to tell you that I was still keeping my cards close to my chest by the time, in August 2012, that I was suspended. This action was quite typical of someone with Reactive Depression.

Now, the suspension itself was a masterpiece of manipulation. A training delegate had made a claim on the Trust for an injury alleged to have been originated on 'a training course' two years previous where 'a trainer', not identified, had demonstrated a police restraint hold after this had been requested by the delegate.

E-mails seen later showed that my Senior Manager seemed to be quite excited about the prospects for this investigation.

However, the Head of Legal Services had interviewed me and told me that she was happy with my story. She sent an e-mail to my Senior Manager stating that no further action was necessary.

But, the e-mail was completely ignored and this incident was used as the justification to suspend me whilst the other investigations continued.

The day after I was suspended, I was subjected to three interviews by three different Investigating Officers. I was offered the opportunity to spread the interviews over several days but I was so punch-drunk that I just wanted to get it all over with.

I am reminded of the words of Julia Roberts in 'Pretty Woman' - "Big Mistake, Huge Mistake!" I was completely unprepared and in no position to give a good account of myself; I gave some ill-considered and inadequate answers to the questions asked by the investigators.

Again, my behaviour at the time was absolutely typical of someone with Reactive Depression; I just wanted to get out of there!

With the suspension, there was no way I could continue to hide this from my family. My wife had been told that her injuries would take 18 months to 2 years to heal. And so, she was very much on the mend physically but my problems came as a tremendous shock to her.

Even then, I only told her about the new allegations not that I was already under a final written warning.

During the following two months, I was not once contacted by one of my line managers. Reviews of my suspension specifically called for in Trust procedures passed by without comment. Only when my Senior Manager decided to go on holiday, was a brief letter sent to me to extend the period of suspension.

The hammer blow came when a date was set for a second Disciplinary Hearing in the October and we were waiting for the evidence pack to arrive. I knew that my cover would be blown then and I could not face telling my wife.

The Slough of Despond

I just did not know what to do, and on the day the evidence pack arrived, I skimmed the reports and the charges against me and hit rock bottom. I concealed the pack until the following day; but, my wife knew that it had to arrive by that day at the latest and she wanted me to chase the HR department to find out where it was. I said I would do this and she headed for work.

It is difficult for me to remember exactly what I was thinking but the closest I can describe is feeling like I was drowning. I was flailing round and achieving nothing and I was still sinking. I remember rummaging around in the garage for some flexible piping and duct tape.

I left the evidence pack on the stairs at home with a note for my wife and then drove off in my car. I headed towards my mother's home about 100 miles away but I could not face her either. I carried on towards the coast but my fuel tank was running low and so I stopped in a lay-by.

I have no memory of what I did next but, apparently, a passer-by saw me in the car with the flexible pipe carrying the exhaust fumes into the car and called the police. They saved my life; I was rushed to a local hospital and woke up the next day to find my mother and her partner by my bedside.

Meanwhile my in-laws had rushed to our home to support my wife who was in shock. They maintain that, over the next few weeks, they became far too familiar with the nuances of the motorway network.

Similarly, my mother and her partner rushed up and down to visit me daily. Thanks to their intervention, I was not released from hospital to go home and they ensured that I was transferred to a psychiatric hospital in an NHS Trust bordering my own. I spent 7 weeks there, initially under strict supervision and then my consultant started to help me to get my head straight.

I was very lucky; the whole family rallied round and started to work on my defence. An employment lawyer was appointed to represent me and a delay was requested for the Disciplinary Hearing.

This did not go down well with my Senior Manager who, in an e-mail obtained later, complained about the delays due to my "illness" (the inverted commas are as used in the e-mail).

Perversely, it transpired that the first hearing date had been arrived at because my Senior Manager was trying to fit in an overseas trip and some time off for family reasons!

But, my illness was never taken into account until the Trust saw one comment from a psychotherapist that said that it would be difficult for me to recover with the Disciplinary Hearing hanging over me. The Trust used that as an excuse to force through the actual date of the hearing.

During all the time I was in hospital and, indeed, all the time since, I was never once contacted by or visited by any of my colleagues.

Contrast this with my wife's boss who is far from a softy but who was extremely supportive. She let my wife take time off when needed; she phoned, e-mailed or texted her every evening to make sure that she was OK; and more than anything else her door was always open if my wife needed to talk to someone

Apparently, someone from HR did telephone a couple of times during the day; but no message was left and I strongly believe that the calls were to check whether I was lurking in a cupboard at home rather than being in hospital.

I left hospital just before Christmas 2012 having been prescribed antidepressants and given a link to my local Hospital at Home team. I was told that I needed to avail myself of psychiatric counseling as soon as possible; however, under the NHS I would probably have to wait three to six months before I could start visiting a counselor.

My family was not prepared to accept this and they made it possible for me to go to a private psychiatric counselor early in the New Year.

Shortly afterwards the trust announced that my Disciplinary Hearing would take place on a Wednesday and Friday in early February; they resolutely refused to delay this any further. As stated before they used the excuse that I would be better off getting this out of the way to justify their decision.

I would definitely have preferred to have waited a little longer but with the right type of representation I felt that I could have attended the hearing. My lawyer asked the Trust to let her represent me. Trust policies provided for this if the individual concerned might be prevented from 'practising their profession' or 'barred from acting in a caring capacity'.

Despite the fact that dismissal seemed a certainty if I could not get my side of this over to the panel, the Trust argued that dismissal did not stop me acting in a caring capacity; only the NMC could do that. So, they refused representation by a lawyer and I had to decline to attend.

We prepared and submitted a well-argued written statement but my lawyer warned me that it would be difficult to control the outcome.

Kangaroo Court

(A mock court in which the principles of law and justice are disregarded or perverted. It is essentially where the defendant has already been deemed guilty, and has little if any opportunity to object or defend himself.)

What we had not quite expected was deliberate manipulation of evidence and procedures in extreme breach of the Trust's own Disciplinary policy and of natural justice with the deliberate aim of ensuring my dismissal.

Early in the New Year we had submitted a subject access request under the Data Protection Act (DPA) to obtain any information on the Trust's databases which referred to me. The Trust quickly announced that they would be taking the full 40 days they were allowed to produce this information.

So, we did not receive the information which showed that one member of the panel had been involved throughout the conduct of the three investigations in coordination, comment, facilitation and decision making.

Not only was she the minute taker and a panel member but, also, she was actively involved in coaching all three Investigating Officers and in the production and approval of all three reports to the Disciplinary Hearing.

Sight of the output from the subject access

request combined with representation by my lawyer would certainly have put the panel on their honour. But, instead my Senior Manager was allowed to introduce all the evidence with little or no questioning from the panel.

As I have mentioned, the hearing was being held on the Wednesday and the Friday and the 'evidence for the prosecution' was completed by the Wednesday afternoon. The panel then moved to consideration of my statement which under their procedures was the hearing of 'mitigating circumstances'.

The Trust's disciplinary procedures specifically exclude the introduction of new evidence at this stage of the proceedings. Clearly, however, my written statement was causing the panel a few problems because my Senior Manager spent the middle day gathering counter evidence to disprove its contents.

I had stated that I had an exemplary record when I was a Staff Nurse in the three years before I joined the training team. This was true and the ward gave me an excellent reference. But, my Senior Manager introduced hearsay comments from people I had worked with and made disparaging comments about one of the references I received when I joined the Trust in 2003 – before I became a Staff Nurse.

Having truthfully said that I had a good sickness record this was denied by my Senior Manager without giving any reasons why.

I also said that I had had 'no problems' at either workplace in the period up to the arrival of my new Senior Manager. Again, derogatory hearsay comments were introduced.

But, alarmingly, HR had also, against policy and procedures, allowed a trawl of their archives; this had produced the report of a ward incident in the middle of 2010. And, by the way, this report was clearly marked 'From the Archives'

A disturbed mental patient in the most dangerous psychiatric ward at the Trust had tried to gain my attention by smashing a TV remote control over my head. The Modern Matron took exception to my attitude at the time despite internal guidance suggesting that allowances should be made during the first 90 minutes after such an incident.

I had received some guidance from my management team at the time but not any formal disciplinary action. The Modern Matron was strongly censured for not providing proper support and for failing to follow trust procedures. Questions should also have been asked about his failure to involve the police in a violent incident against a member of staff.

When I used the term 'problems', I was referring

to disciplinary problems but perhaps I should have been more careful about the way I expressed myself. But then, you don't really expect HR managers to risk a charge of gross misconduct by revealing spent personnel file information to help business managers to achieve their objectives.

My Senior Manager was obviously also worried about my having explained clearly my foolishness about the MRI appointment for my wife. The fraud investigator was contacted to find out whether an appointment with someone else with the same surname existed for that day.

As nothing further was said about this at the hearing, one can only suppose that the investigator confirmed the existence of my wife's appointment.

At no point on the middle day or, indeed, throughout the three-day period was I contacted for clarification and, obviously, there was no opportunity for me to give my version of events.

Inevitably, with no one to present my case or counter the arguments presented, the panel found against me.

On the back of hearsay evidence and evidence which would never be allowed in a court of law, all the arguments for mitigation were overturned.

The panel cited a 'number of problems over a number of years' that had 'been addressed with' me 'informally by previous managers' in rejecting those arguments.

Not one of those previous managers was called to give evidence; my Senior Manager was allowed to quote them without having to provide any proof.

Within minutes of the success at the hearing, someone at the trust arranged for the cancellation of the insurance on my Trust lease car. This must have been their equivalent of a fanfare of trumpets. However, I only received notification of my dismissal and the ending of the car lease a week later. Oops!!

The Truth Will Out

Our DPA subject access request was submitted to the Trust on 22nd January and we should have received all the output by 3rd March. This was inconvenient because we were due to confirm our Appeal to the Trust by the end of February; by that date we had not even received the minutes of the hearing.

My lawyer spoke to a new, more reasonable person in HR who agreed that we should not have to notify our appeal until the DPA output and the minutes of the hearing had arrived; an extension of two weeks was granted and my lawyer was given leave to represent me at the Appeal.

E-mails, seen subsequently, show that my Senior Manager was extremely upset by these concessions and the divisional head had been asked to intercede.

The HR representative made strong and successful arguments for both decisions; but, in doing so, she effectively confirmed that my lawyer should have been allowed to attend the Disciplinary Hearing let alone the Appeal.

We finally received five lever-arch binders and two computer discs containing e-mails, reports and other paperwork relating to me on 7th March. A quick examination showed that we had received some e-mails for 12 Trust managers but it was clear that some of the e-mails even for my Senior Manager were missing.

No e-mail attachments had been provided and a number of key documents like the written complaint from my female colleague were absent.

No e-mails had been provided for 13 people actively involved in investigations against me and a further 32 people who had been involved in e-mail conversations had been excluded.

And yes, that does mean that some 57 people had been involved in all the activity against me over a period of 2½ years; There seemed to be no rhyme nor reason to some of the involvement; people were being recruited to help on an ad hoc basis.

It seems to me that there was no formality to this and so my managers had no idea what all this was costing; there was no incentive to manage properly, abandon trivia and focus disciplinary activity only on serious concerns.

Surely the NHS has a duty to taxpayers and should funnel serious disciplinary investigations through a separate specialist unit that charges out its time to the commissioning manager.

The minutes eventually arrived with my lawyer on the 11th March and the Appeal deadline was extended to 18th March.

But, we then found out that, rather than proper

minutes, we had illegible, incoherent, handwritten notes of the hearing. It was as if someone had said that the outcome was the end of things, I had finally been got rid of and the minutes were of no importance. Frankly, I think this was shameful.

On the 11th March my lawyer pointed out to the Trust that half of the data from the subject access request was missing

The typed minutes were received by my lawyer on 27th March and, in fact whilst legible, they were the same garbled, incoherent notes that by then we had deciphered with six of us working together and sometimes using magnifying glasses.

So, these minutes added nothing to the sum of our knowledge and they certainly did not reveal any incisive cross examination of any of the witnesses for the 'prosecution'.

In common with the statements prepared by the Investigating Officers, the witness statements given at the hearing were light on facts, times and dates and heavy on supposition and hearsay.

If something might have been true against me it was true; if something might have been true for me it was untrue; the evidence from the course delegates which cleared me of making a fireman's lift was completely ignored.

On the 28th March, over 2 months after our DPA subject access request and over two weeks after my lawyer pointed out that a lot was missing from the first output, a letter from the Trust asked for clarification of our requirements and suggested that some people had refused access to their e-mails for confidentiality reasons.

Now, any organisation worth its salt has a Data Protection Officer (DPO) who knows exactly what should be provided under a subject access request and also knows that no one can refuse access to relevant information. And, most DPOs would be mortified if they could not complete their task within the 40 days allowed; most have access to the Chief Executive if obstructed in their role.

I am quite confident that the Trust knew exactly what to do but it also knew that, the longer it delayed providing the information we asked for, the less time we would have to analyse it.

This latest letter also gave us one week to submit our Grounds of Appeal but we had been hard at work and submitted them the same day much to the Trust's surprise.

My lawyer also pointed out that, if the Appeal was not scheduled before May 7th, we would have to submit a protective claim to the Employment Tribunal because with all the delays by the Trust we were up against another time limit.

The Trust came back with an offer of an Appeal Hearing at the end of the first week in June with the excuse that not one Director was available for an appeal before then. This is a Trust that in its latest edition of its Disciplinary Procedures states that any appeal must be held within 20 days of the Disciplinary Hearing.

Needless to say, we submitted a claim to the Employment Tribunal on 7th May and included the failure to provide information and the delay in holding the Appeal in our evidence.

A friend described the behaviour of the Trust as 'Like a 500-pound Gorilla on Speed'. The organisation seemed to relish the use of its bulk and its power but what it was doing was just plain stupid. Withholding information just raises suspicion, delaying is both irritating and unfair and pushing people to the brink usually results in a hard push back.

My lawyer finally received the rest of the DPA output just one week before the Appeal hearing. There were some useful snippits to add to her cross-examination briefing but, frankly, nothing that made a fundamental difference to our case.

I was lucky; I had the financial, emotional and intellectual resource of my family behind me. The combination of my lawyer, my wife who is a very senior executive, and the rest of the family comprising a retired Chief Executive, a former Banker, a retired Chief Superintendent of Police, a Nurse Practitioner and a Senior HR Manager made for a formidable team to examine all the paperwork, to review the Trust's policies, to draft briefing papers and to prepare evidentiary submissions.

One of the interesting products from all this work was a detailed analysis of the huge number of breaches and misinterpretations of trust policies revealed by our work. Many of these breaches could result in charges of gross misconduct against the person responsible. So far, no such charges have been forthcoming. However, my family has vowed to try to ensure that the truth comes out.

The Appeal

Sometime before the Appeal hearing it was announced that the panel would be chaired by a Non-Executive Director of the trust supported by two senior managers. This is despite the fact that the Trust disciplinary procedures in force at the time of the investigations stated that 'Appeals against dismissals will be considered by a panel of three Directors at least one of whom shall be a Trust Board Non-Executive Director.'

The latest edition of the Trust disciplinary procedures, surprise, surprise, allowed the panel that was selected. So, a retrospective rule change was acceptable here but not to ensure an Appeal Hearing within 20 days! The Trust seemed to choose the rules that fitted their purpose.

One of the senior managers was a close colleague of my Senior Manager and reported in to the same divisional head. We protested but were informed that that was how it was going to happen – so there!

My lawyer asked whether it was possible for one of my relatives to be present to support me because of my illness. No - it was not; we had got my lawyer into the room and that was presented as a big concession even though the Trust had already reported me to the NMC with a view to stopping me practising.

But, to be fair, a very typical NHS office, with DIY melamine covered shelves fixed at dangerous heights and that shabby-chic look that we all know and love, was put at our disposal.

Much to the Trust's surprise, my father-in-law and my mother's partner set up camp in there for the day. You have no idea how good it felt to have a 'siege party' outside the walls. To be fair no one put any obstacles in their way.

The next revelation when my lawyer and I entered the room was that the Management Representative was the Chairman of the Second Disciplinary Hearing and not my Senior Manager.

As my mother's partner, the former Chief Superintendent, quickly pointed out, that is the equivalent of the Judge who presided over a High Court case acting as the Prosecutor at an Appeal Hearing – morally unacceptable, completely out of line with normal court procedures but alive and well in my NHS Trust.

In a prominent supporting role was the HR manager who had been involved throughout the conduct of the three investigations, who was a panel member and who took the minutes of the second Disciplinary Hearing. I guess she had to be there because she was, after all, the only person who could really understand her notes of the meeting!

Mind you, this time, the Chairman of the panel of three had someone who was described as 'Secretary to the Panel and HR adviser' plus a separate note taker on hand. Yes, that is 7 people to our 2; mob handed or what!

So, the Chairman of the Second Disciplinary Panel was allowed to spend most of the morning reprising the evidence and justifying the way she had handled the Hearing including the reasons for allowing new evidence at the wrong time, the reasons for accepting hearsay and unsubstantiated evidence and why I was, thus, regarded as untrustworthy and deserving of dismissal.

She was never questioned about the failure to ask searching questions of the witnesses or the failure to recognise evidence that was in my favour.

In our Grounds of Appeal document, we had reminded the Appeal Panel that the trust's Disciplinary Procedure specifically states that 'No statement of previous acts of misconduct by the employee or the issue of a Formal Warning or Warnings unrelated to the alleged offence(s) on which the Disciplinary Action is based, shall be made'. Needless to say, this was ignored.

Rightly, my lawyer decided to give her no more air time and moved straight on to cross-examining my Senior Manager. Having embarked on an extensive grilling for some three-quarters of an hour it was obvious that my lawyer was getting under my Senior Manager's skin.

But, a straight bat was played and the panel failed to follow through with any searching questions about perceived management failings and policy breaches.

When my Senior Manager was asked about the failure to pursue an informal resolution of the Harassment and Bullying allegation, the excuse was that this was to avoid upsetting my female colleague any more than she already was. The fact that I was sufficiently upset to want to take my own life seemed to be of no importance.

When asked for a brief summary of her management philosophy, the reply was 'Firm but Fair'!

The next witness called was my junior line manager but he was not there. Despite, my lawyer's request for him to be present, someone had told him that he probably would not be wanted and he was at another location carrying out a training session.

The Chairman of the panel was not at all happy about this and my line manager was ordered to present himself for questioning without delay. But we were already late into the afternoon and the loss of nearly half an hour was not fair to my case.

When he arrived, my lawyer did give him a hard

time and he did not make a good witness. Initially he denied that my female colleague had spoken to him during the week of the course we taught together; this was despite her having given evidence twice that she had spoken to him.

And, he used that statement to excuse letting us sit together at the Team Away Day. Later, he actually admitted that she probably did speak to him.

Asked about following the guidelines on informal resolution of her complaint against me, he virtually admitted that he was too scared to do this and he had immediately passed the whole thing to my Senior Manager.

So, despite being our line manager he abdicated responsibility for his Duty of Care to both my female colleague and to me. For the life of me, I cannot see what useful purpose he served other than to put another layer of protection between the 'workers' and my Senior Manager.

All he succeeded in doing was providing my Senior Manager with what must have been regarded as the final piece in the disciplinary jigsaw. And, clearly any attempt at informal resolution might have resulted in the loss of that glorious opportunity.

My lawyer did a good job of summing up and then I was allowed to make a personal statement.

If I had had to do this a few weeks earlier I would have cracked up but I did it and my lawyer said that I did it well.

The HR manager assisting with the Management Case tried to interrupt me to dispute something at one point but, the Chairman of the panel put her in her place and allowed me to finish.

In the end, it could not make a difference. To make a concession to me, the Trust would have had to admit that I had been suffering from Workplace Stress and Reactive Depression during the whole period covered by the various investigations and hearings.

To do that would be to admit to a level of liability which could be very costly.

The panel eventually judged that, regardless of the fact that some of the accusations were trivial and that the proof in certain cases was thin, the totality of the misdemeanours was sufficient to justify dismissal.

They said that they had considered whether I could be employed in another role but felt that they could not recommend this because of the 'trust issues' involved.

And, as I have readily admitted, I shot myself in the foot on several occasions but that is an explicit behaviour of someone with Reactive Depression.

Where I believe that the panel over-stepped the mark was to suggest that they did not believe that my Senior Manager's behaviour had been inappropriate.

And, another manipulation was to suggest that 'the timescales were satisfactory' using my illness as the excuse.

They should have, as a minimum, qualified both those observations.

My family encouraged me to pursue the Appeal as an exercise in rehabilitation and frankly, from the previous behaviour of the Trust, we were not expecting a different outcome.

Right up to the week before the hearing I was very much in two minds about attending and, if someone had let me off the hook, I would probably have been very glad.

But, by the weekend before the hearing and thanks to huge encouragement from my family and my psychiatric counselor, I gained a new determination.

And, I think the Appeal was good for me; I could banish my demons; I faced up to my 'foes'; and, I spoke firmly and convincingly in public for the first time since my breakdown.

I pulled no punches and, if nothing else, my Senior Manager knows that I think that the Harassment and Bullying allegation was made against the wrong person. And, perhaps just perhaps, that may have stirred some feelings of remorse. But, I am not holding my breath!

The Last Word

This book has not been written out of bitterness. The first purpose was to allow me to tell the whole story as I see it. And, every word I have written can be substantiated and proved in depth thanks to the huge volume of information extracted from the Trust in our DPA subject access request.

The second purpose was to try and get an understanding at large of the appalling failings in staff management in parts of the NHS. It is a punishment culture not a caring, coaching and training culture. Disciplinary action tends to be a first resort rather than a last resort as it should be.

I know now that my Reactive Depression caused me to make many, silly mistakes; and, I do regret the pain and anguish that I have caused my family. However, my family has been nothing but supportive

This is a very direct contrast to the failings of my NHS Trust. If psychiatric illness in a nurse employed in the NHS can be completely ignored by qualified clinicians, how can patients hope to receive the help they need before it is too late, which it nearly was in my case.

And, how can a supposedly caring NHS Trust and its supposedly experienced managers become so obsessed with written rules, processes and procedures and yet habitually ignore or pervert so many of them? Surely, that is the sign of a deeply troubled and unhealthy organisation in need of root and branch surgery.

Anyway, I have left the **NHS**, I certainly won't be needing that **New Uniform** and they took my **Blackberry**® away when they suspended me!

Conversations with the Family

Whilst writing 'The NHS, the Uniform & the Blackberry®', one thing that became more and more apparent is that James has a formidable fighting force behind him, his family. In my opinion, for him, that has been, quite literally, the difference between life and death.

So, I decided that the story would not be complete without getting some insight into their feelings and motivations. What makes an extended family with different backgrounds unite in such a determined and forceful way?

My first interview was with James's wife, who is a together, intelligent woman who holds down a challenging senior executive role. After gaining a good degree, she completed her doctoral studies and spent some time in academia. Now, she has a broad set of responsibilities with quite significant resources and budgets.

She had known James since he was nearly 14 years old. She first met him when he was with 3 friends on a PGL skiing holiday. They were immediately smitten with each other, remaining boyfriend and girlfriend for some 2 years.

But, they lived a fair distance apart; and neither of them drove a car; and, they were both working towards exams; so, inevitably they drifted apart.

Over the next few years they both got on with their lives but reunited whilst she was undertaking her doctoral studies and running a local wine shop to help with the finances.

James had been working for some time in the NHS but had no relevant qualifications. Their renewed relationship encouraged him to settle down and get qualified.

Five minutes into our chat, it was quite clear that this was also one very angry woman who, at times, has great difficulty dealing with the emotions which have been stirred by the ill-treatment of her husband.

As a senior executive in a well-run business, she was horrified by the callous, if not promiscuous, use of formal disciplinary procedures in the NHS Trust.

And, having had the importance of rules and regulations drummed into her by her parents, she could not understand how so many rules could be written only to be ignored or perverted at will.

Clearly, the past few years have been a nightmare for her because, as the injuries from her riding accident faded away, something akin to 'Armageddon' came along and assaulted her senses.

She had known that James had become unhappy in his job since the arrival of the new Senior Manager but nothing had prepared her for the story which unfolded in October 2012.

Initially, her anger was directed at James because she had so few facts to work with. But, being someone who was used to working with a lot of random and complicated information, she soon put all her energies into finding out the truth.

Early on the scene, after 'Armageddon', was her Mum. Her first comments would probably be classed as typical of a mother-in-law. She said that her initial reaction was that she could have cheerfully strangled James and suffered the consequences.

It seemed to her at the time that he had put his wife and family through so much over the past few months. However, she had no idea what he had been going through or, for that matter what was going to hit the family over the next few months

For nearly three hours on the road rushing to her daughter's side when James disappeared, she had worried herself silly.

Memories were stirred of a previous tragic time when James was a teenager. During the 2 years following that, James had joined their family for several holidays. And, she recalled that he was often a bit like a puppy – big paws and very energetic, but always fun to be with.

To use her words, 'after leaving school, James had bummed around a bit without finding direction'. Then he had become a care assistant in a psychiatric unit, married their daughter and settled down to gain his nursing qualification.

She was quick to say that James is a very caring person and had always looked after her daughter. Ruefully, though she realised that he, perhaps, had cared so much that he failed to seek help from his wife or from the family when he really needed it.

She recalled that when her daughter had had the accident, falling from her horse, she was, to not put too fine a point on it, very lucky to be walking. She had had the 'privilege' to view her back shortly after the accident and it was a mess.

With hindsight, she recognised that the riding accident, the investigations at work and the death of the granny he had relied on in his teenage years all came at the same time.

And, of course, the fact that he was being systematically bullied and harassed by e-mail and in person only became apparent after the event.

She was obviously emotional about the fact that James had felt that no-one could help him. But, she realised that this was all part of the Reactive

Depression that he had suffered from over two years, a gradual process that had finally worn him down.

----- ◆ ◆ ◆ -----

At the same time, I met the man that James often calls 'Boss', his wife's father. He is an experienced businessman who retired some years ago after 35 years with financial services organisations and 10 years running a multi-million-pound international business.

Whilst his daughter recovered her composure, he had acted as the driving force behind James' defence. He readily admits though that he had to quickly move to one side once she had 'got a grip'.

He expressed his liking for James despite everything and it was quite clear that there existed a real friendship of opposites.

As a very focused, determined and, he admitted, a very aggressive person at times, he just could not understand how anyone who had significant business responsibilities could spend more than two years and a considerable amount of taxpayers money pursuing a vendetta against one of their subordinates.

And, he believes that the way that line managers are allowed to pursue their own investigations, using any spare member of staff available, is bound to lead to team bullying.

If managers had to bear the cost of experienced third parties carrying out their investigations for them they would not pursue trivial issues just for personal gratification.

Yes, he had had to make people redundant sometimes but that was part of the job. It did not have to be done cruelly and a good company put in place a welfare safety net and proper financial compensation to see the victims through.

As a great believer in team dynamics, he had also occasionally had to get rid of people who failed to fit or, over time, became an irritant within the team.

Usually though, the solution was to coach those people and encourage them to move on. There were times they had to be nudged but, again, this could be done kindly and with a suitable financial incentive.

During his entire time as a Chief Executive, he had maintained the rule that he was the only person who could chair an Appeal against dismissal arising from a Disciplinary Hearing. And, he had never once had to do that.

He showed a deep contempt for a comment made by the Chairman of the Second Disciplinary Hearing in her written Management Case to the Appeal. She said, 'I am very experienced at chairing disciplinary cases'. He said, 'that shows something is wrong with the culture in that NHS Trust'

In his organisation, it was a culture of coaching, formal training and regular appraisal that produced a workforce and a management team who performed well and disciplined well.

He could not stomach the idea that an organisation could bully, harass and, using his words, 'torture someone to the brink' and over such a long period when so many alternatives were at their disposal.

As someone with a fairly good understanding of business costs, he had made a conservative estimate, ahead of the Appeal Hearing that the NHS Trust in question had spent anything upwards of £150,000 on this process. And, if you took into account lost productivity and over-manning to cater for this type of disciplinary activity, the total cost could easily reach £250,000.

As a taxpayer, let alone a father-in-law, he was not amused.

I moved on to meet James' mother and her partner.

His mother, an experienced Nurse Practitioner,

has clearly had more than her share of family traumas over the years and recently.

She readily admitted that her husband's premature death, whilst an appalling tragedy for her, came at a terrible time for James. He had worshipped his Dad and they spent a lot of time together doing 'boy's things'.

His Dad's death and her understandable withdrawal into herself left a void in James' life which was only partially filled by his grandmother.

James found solace with his skate-boarding friends and their families and for a while he did become a bit feral. And, on reflection, at that time he had shown a tendency to keep his thoughts to himself

But, she had been proud of him for getting his qualifications and had watched him develop into a caring grown-up with some real satisfaction.

With 20:20 hindsight, she realised that he had reverted to his teenage years when confronted with the hammer blows of problems at work, his wife's accident and his granny's illness. He had just absorbed the punches and had spoken to no one about his feelings.

This was where her anger with the NHS came out. Everyone working in the NHS knows that the standard of management in some Trusts is appalling. Very few senior managers know anything about coaching or encouraging their people.

The tendency is to use formal disciplinary procedures for the slightest infraction and suspensions are, in her words, 'handed out like sweeties'!

She mentioned a National Audit Office report 10 years earlier which had highlighted the more than £40 million a year that was wasted then on suspended doctors.

This report had identified part of the problem but she wondered what the true cost was of all disciplinary procedures and all suspensions of doctors, nurses and ancillary staff in the NHS.

When she had first become a nurse, the NHS had been a great place to work. But, the leadership from the very top had become out of touch and insensitive.

Bureaucracy was rife and her feeling was that 'more money is spent on secretaries and subsidised cars and in writing procedures than in looking after patients'.

She admitted that, from being a great enthusiast for the concept of the NHS, she now doubted it could continue to exist in its current form.

Her partner, a very tall, amiable man with a

twinkle in his eye, then surprised me with the authority and the vehemence with which he spoke.

As a retired senior police officer, he could neither contemplate nor accept the way the trust had carried out its investigations and its hearings. Everything about this was out of line with police and court procedures and, as far as he was concerned, a flagrant breach of natural justice.

For the second Disciplinary Hearing, two of the three investigators had never done such an investigation before. And, throughout, they were being actively coached by an HR manager who then sat on the Disciplinary panel.

The evidence, such as it was, lacked substance, dates, times and corroboration. Where evidence was inconvenient, such as the statements by the delegates who denied seeing a fireman's lift, this was pushed to one side.

Most of the allegations were trivial issues which in a well-run organisation would have been dealt with by day-to-day management interaction.

Even the fraudulent alteration of the MRI appointment letter could have been dealt with compassionately if James' managers had been interested in what was happening in the lives of their staff.

He had, of course on occasions, had to pursue formal disciplinary action against his men but not very often. And, when he had, this was immediately taken out of his hands and passed over to another division. He neither investigated, prosecuted nor had any role in the outcome.

It was anathema to him that James' Senior Manager and the HR adviser allocated to that department could have been the motive force behind the investigations and then turn up at the hearing as the Prosecutor and one of the 'judges' respectively.

For the Chairman of the Disciplinary Panel or, in his words, 'the Judge' to turn up at the Appeal as the prosecutor was completely unacceptable.

In his opinion, the NHS is a big enough organisation to adopt the same approach as the police and divorce disciplinary investigation from the line managers originating the complaint.

And, throughout this, as someone who had had to work to the letter of the law, the blatant disregard of internal rules and procedures whilst driving a man to attempt to commit suicide produced a volcano of emotions which belied my first impressions of this man.

In the police force, the twisting of evidence to secure a conviction of someone innocent of an offence but believed to be 'a bad egg' was called 'noble corruption' and he felt that this label could readily be applied to James' managers.

A final blast of frustration produced the wry

comment that all this deception and the repeated perversion of rules and procedures smacked of a combination of the oppressive regime of Joseph Stalin and that of Big Brother in George Orwell's '1984'.

However, on further reflection, he thought it all seemed more like Lewis Carroll with the Red Queen in 'Alice in Wonderland' saying "No, No – sentence first, verdict afterwards" or Humpty Dumpty in 'Alice Through the Looking Glass' saying "it means what I choose it to mean – neither more nor less". And, he reminded me of what had happened to Humpty Dumpty!

My last visit was to see James' brother, a senior HR executive in a large multinational organisation.

Put simply, he was dismayed at the unprofessional way that the HR managers at the NHS Trust had acted. Indeed, he felt that the Chartered Institute of Personnel Managers (CIPD) would certainly censure at least one of them and possibly more.

His view was that an HR manager is there to advise and support but that that applies to both managers and their staff. They are not there to act as line managers and they should not take sides in disciplinary investigations or hearings.

He knew that his brother had 'been a bit daft' on occasions but also was astounded that, of all places, the NHS could not pick up the symptoms of depressive illness.

More and more organisations are being compelled to identify and support disability and mental illness. As far as James' brother is concerned, the first place you would expect this is in an organisation that has to care for disabled and mentally ill people.

Like the rest of the family, he was completely disillusioned about the current state of an organisation that used to be so well respected all over the world.

For the whole family, the whole experience had not been easy. But, all of them put their collective brains together and became a force to be reckoned with.

If nothing else, they have shown the Trust how incompetent and inefficient they are. There may be Policies, Procedures and Directives for every possible eventuality but they are ignored at will.

The provision of material support which James' Senior Manager and line manager were so proud of was regarded as irrelevant by the family.

They wondered where was the humanity, where was the compassion? All they saw was constant criticism and a tendency to admonish and investigate anything that moved.

Totally missing, in their view, was guidance, encouragement, and, heaven forfend, praise for a job well done or a qualification attained

Book 2

Justice is a Long Road

<u>The Way Back</u>

My immediate thought was that the last thing I wanted to do was to go back to work in the NHS. But, I knew that I enjoyed caring for people and that I was quite good at it.

And, my family persuaded me that I should not burn my boats as I might want to work my way back into the business of care.

Initially, I took up two voluntary roles. One was with a neighbourhood befriending group nearby, taking older people shopping and providing help in the home. Eventually being asked to appear in a video to promote this scheme was a big boost to my morale.

The other role was at our local zoo; no, this is not on the grounds that I am an expert with 500-pound gorillas but because I love animals (except bats!); and, a breath of clean fresh air is something I rarely got in my NHS job.

I got a real boost to my confidence from this job too when they spotted my communication and training skills and moved me into their education area.

My family and my therapist continued to encourage me to take things slowly and not to make any rash moves. I could see their point and knew I needed to be very careful who I worked for in future.

Eventually, a year after my suspension from the NHS Trust, I started a new job in an entirely different business with absolutely no links to healthcare. I found that I enjoyed the new role and the new routines; and, imagine my delight when I discovered that I liked my new employer.

The nice thing is that my managers are generous in their praise and appreciative of my skills. There is no lack of discipline but I can truly say that they really are firm but fair!!

Their induction training has been exemplary and I came out of the initial probationary period with flying colours.

It was at this point that I made the final decision to abandon my old career and prove myself in this new one.

So, I have to start from the bottom but I have found a workplace I enjoy, an employer who cares and colleagues I get on well with. It may not seem much of a career at present but my manager put my name forward for a management training course which I am close to completing.

One thing is certain; my past experience has given me excellent guidance in how not to manage people. The management at my NHS trust showed themselves as past masters at that!

Most importantly of all, for the first time in many years I now know what it is like to get up in the morning, go to work and be very happy. I am well aware that there will be obstacles ahead but I have a much more positive outlook now. And, with help from my family I think I can surmount most of those obstacles.

NMC Referral

I knew that my NHS Trust intended to refer my case to the NMC. However, having told the Trust that I intended to Appeal you would think that they would await the outcome before doing anything more. Think again!

No sooner than my Disciplinary Hearing was over, my Senior Manager eagerly rushed into preparation of the submission to the NMC. Within a couple of weeks, all 350 odd pages of evidence was bundled up and submitted. This was obviously regarded as the final coup de grace.

In their customary efficient way, the NMC did not let me know about this until 16 weeks later, in July 2013. They had not done anything in that time except log the complaint but their procedures say 16 weeks and so 16 weeks it is. I have become convinced that people at the NMC are controlled by a precisely calibrated, timed, computerised progress sheet!

Come August, almost exactly six months after the Trust submission, the NMC sent me a copy of their documents which were devoid of any of my arguments or statements or, indeed, anything from the Appeal. That was, of course, inevitable because the Appeal hadn't happened when the Trust made their submission!

They asked whether I had anything to say in return. I certainly did and I sent a 400-page bundle including all the missing paperwork and all the arguments my lawyer and I had provided to the two hearings.

My advice to anyone else in the same position is don't bother to do this. The NMC does not look at your submission at this stage and does not even look at what you have sent them when they have compiled their evidence against you. My wife swears that they shredded it all.

Instead, the NMC, at great expense, hired a firm of lawyers to take all the evidence once again from people at the Trust who had previously given testimony against me. And, this was even though I had not disputed some of the allegations.

I was claiming that I was ill and behaving in an uncharacteristic way because of the bullying I had endured; I was not claiming that I was completely innocent!

I know that they did, in fact, interview at least one former colleague who spoke well of me. That evidence was missing from the 1,249 pages of documentation which was sent to me in January 2014 ahead of an NMC Investigating Committee meeting set for the first week in February.

Not only was any positive evidence missing from the investigation report but I found out later that the investigators had not been allowed to see any of my submissions; and, they were not even made aware of my health problems. You might forgive me for thinking that the NMC were deliberately stacking the cards against me.

It is perhaps stating the obvious but the date of the committee hearing would be exactly one and a half years from the date I last worked at the Trust and one year since I was dismissed. The NMC seems quite content to put practitioners lives into a sort of suspended animation for inordinately long periods of time.

I ask myself what would happen to a suspended nurse who was ill with depression, had tried to commit suicide, still had a partner and children to support and did not have the benefit of a well-educated, supportive and determined family?

Well, of course, the financial, family and emotional pressure would be intolerable. Research indicates that a second suicide attempt would ensue; and, this time, it would probably succeed.

Luckily, I do have a very supportive family and I had a good psychotherapist (outside the NHS) who saw me through the dark days of feeling unwell and unwanted.

I had the good fortune to have time to rethink my situation and decide whether I wanted to risk returning to a career for which I had trained for many years and which I had loved. But, I had finally recognised that, with bad Trust leadership and the wrong working environment, it was a career that had put my life in jeopardy.

So, having established a new career and finally understanding that a return to the NHS would probably be the last thing I should do, I decided optimistically that the time had come to shut down this episode in my life. My optimism was misplaced!

The NMC asked whether I had any comments for the Investigating Committee. I sent them a letter indicating that my submission in August 2013 should be taken as my final rebuttal to the accusations. No-one told me that they would not accept that and that I should have made a full response to the allegations at that stage

I asked them to take account of my illness; and I explained that, having visited a psychotherapist for nearly a year and taken daily medication, I now realised how unwell I had been and for how long I had been unwell.

I expressed my regrets at what had happened. But, I also drew attention to the vindictive actions of my Senior Manager and the unprofessional treatment I had received from managers and colleagues. Their failure to take note of the many years of good work that I had put in at the NHS and the fact that I had never harmed a patient showed a very shallow approach to man-management.

There is no doubt that the way in which multiple investigations, some justified some not, were stretched out over two and a half years directly exacerbated my underlying illness and led me to such feelings of desperation that I tried to take my own life.

This was compounded by a complete absence of constructive management interaction leading to a feeling of abject isolation. Frankly, I felt that I was working in a toxic environment lacking humanity and compassion.

In a team of qualified nurses and mental health nurses, surely it could be expected that someone would have spotted my developing illness.

Lastly, I indicated that I was happily engaged in a new career, away from healthcare and that, at that point, I had no plans ever to return to any job in healthcare. I said that, should the panel decide that my case needed a hearing, I would be unable to attend as I had no wish to jeopardise my recovery. I stressed that my continuing good health was the first priority for me and my family.

In mid-February, just one week after the NMC Investigation Committee met, I received a letter from the NMC. This time they enclosed a 20-page questionnaire for completion. The questionnaire asked in detail for my analysis of and response to each of the perceived offences. And I had to fill in, yet again, a load of personal information.

Of course, all this information had already been provided in the evidence pack I submitted the previous year. And, I had already told them my response to the allegations.

As a bonus, I was invited to go to the website and download another 6-page form if I wanted to apply for Voluntary Removal. But, I had already done that in my letter!

A case worker is allocated to each NMC investigation but all they seem to do is receive a prod from the computerised progress system at set intervals and send standard letters and forms.

They don't read and understand submissions and never précis anything for the committees. Why? Well, of course, the 20-page form does this job for them; but, why didn't they send that out in the first place?

As an added bonus, without consultation or explanation, I was handed over at this point from a Case Officer with a name similar to that of a Jedi Knight to someone with a name similar to a Russian Cosmonaut. To me, the NMC had shown itself to be as sterile as my former NHS Trust and just as inconsiderate and unfeeling.

At no time prior to the Committee meeting had anyone at the NMC made any allowances for my illness or modified their highly bureaucratic and officious standard letters because of it.

But then, if you visit the NMC website, there are no concessions to disability or illness, whether physical or mental, despite the requirements of the Disability Discrimination Act. Clearly, the NMC insists that all its registrants remain in rude good health for the complete duration of their extraordinarily protracted investigations!

My initial and, I admit, irrational and angry inclination was to do what my parents did with Readers Digest publicity many years ago - stick all the paperwork back in the envelope and return it without a stamp.

Then I broke into a rant going something like "I WANT TO BE LEFT IN PEACE, I DON'T WANT TO REVISIT THE NIGHTMARE, MY LIFE HAS MOVED ON, WHY WON'T THEY LEAVE ME ALONE; DO THEY HAVE ANY IDEA WHAT THEY ARE DOING TO PEOPLE? I WANT TO BREAK FREE!!"

Before I could launch into a dodgy rendition of that Queen pop song, my wife came along, scraped me off the ceiling and then mopped my brow. "There-there she said it will be alright".

She took away all the paperwork and gave me a cup of tea. Tea is a wonder drug for people like me. It soothes the fevered brow and makes the world seem whole again!

A little later I heard mutterings from the study. "This organisation seems to be like the Catholic

Church; you confess all your sins, say a couple of Hail-Marys and you are eventually forgiven!" After a short interval, there came "Why don't they provide everything on their website so that you can complete and submit the forms on-line like you can for any other grown-up business? Do the NMC live in the dark ages? Every government agency allows you to do this - why not the NMC?"

Then I heard a muffled scream and "You can download the Word version of the Voluntary Removal form but it's not set up in the right way for completion on screen. What moron created something which has to be printed out and completed with a quill pen dipped in blood?"

This was quickly followed by "And, they send a silly 20-page paper form with tick boxes because their people are not prepared to examine and précis the hundreds of pages of information they have already been provided with.

But, they are happy to spend a fortune on lawyers to confirm the Trust's view of reality. And what cretin decided that this form should be sent by post rather than provided for download on-line? They obviously don't give a stuff about wasting the nurses' money".

Yes, I am the one who has had mental health problems but I could hear someone else quietly becoming unhinged. Better to keep out of the way for the moment perhaps?

The next thing I heard was my wife picking up the phone and dialing. She was phoning her Mum to share her frustrations. After a short rant, I heard the first giggle and then they started to discuss a hilarious plan which involved hand grenades, rocket launchers and concrete overshoes.

Whilst they were planning their campaign I reflected on the attitude of the NMC to my illness. Registration and regulation of nurses started in 1860. At that time, someone with mental health problems would have been confined to a sanatorium or marginalised to a workhouse for the rest of their life. Whilst attitudes amongst the public at large may have changed, the NMC seems still to have a Victorian attitude to mental illness.

Why were they afraid to recognise my illness? Why can't they tailor their operations to help someone who is recovering from mental illness?

Would it undermine their credibility if they were to ask my NHS Trust why they failed to recognise my illness? Is it unreasonable to expect them to identify the failings of the NHS Trust management and their poor attitude to staff relations?

The truth is that, whilst the NMC does need to be there for complainants, it has actually become a creature of the NHS Trusts. It has forgotten that the nurses and midwives are the paymasters and should be calling the tune. Perhaps the time has come for the NMC to become accountable to its paying registrants and subject to formal annual audit of its processes and procedures?

Whilst NMC registrants continue to meet all the costs, and nurse and midwife registration fees have increased massively in recent years, the Trusts can refer staff willy-nilly without bearing any of the cost. It does seem a bit much to be put in front of the firing squad whose wages you are paying and whose guns you are buying and then to have to fork out for the bullets!

The NMC Chief Executive claimed that fees would have had to have risen a lot more if they had not 'achieved efficiencies'. This assertion is unlikely to be greatly appreciated by hard working nurses trying to cope with their other outgoings.

If Trusts had to pay an up-front fee or a fairly substantial deposit for referrals, they might have to manage better and treat their staff like human beings. It is, after all, usually poor management, management weakness or inadequate training that leads to staff making mistakes or behaving badly.

And, whilst I am no whistleblower, with an up-front payment for each Trust referral there would be a disincentive to using the NMC as a bludgeon to cower or subdue whistleblowers.

Having reflected on that, I realised that I could be of most use if I put the kettle on and made both of us a cup of tea. By the time I delivered the magic brew, some serious discussions were underway between mother and daughter; they had agreed that 'Team Family' needed to get into letter drafting mode.

The 20-page form was irrelevant because they already had all the information in their files. It was decided that I would complete and sign the 6-page Voluntary Removal form but it would just refer to a letter written by my wife and ask for all future correspondence to go to her.

Some would say that I was running away but I had suffered 3 years of mental anguish caused by bullying, mobbing and ill-conceived, vindictive, vexatious management behaviour. That was enough without having déjà vu all over again! I was actually well on my way to recovery and starting to live and enjoy a normal life

My family, bless them, prefer me as I am now. My father-in-law said very recently that at last he could have a lively conversation with me without my shutting off and going into my shell. I and my family would prefer me to remain sane and alive.

My wife decided to throw her current thoughts into a Word document. This would then be subjected to a round robin process of editing and evolution until it was carefully crafted and more like a surgeon's scalpel than an artillery barrage.

The NMC needed to be in no doubt as to our decisions and the fact that I was taking control of my life. We were angry but businesslike enough to know that our words had to be chosen carefully.

The final version of her letter, sent on February 23rd, expressed concern about the impact that dealing with the NMC was having on my health. She asked them to have no further contact with me and this was confirmed by me on the request for Voluntary Removal from the register which was enclosed.

She expressed surprise that she had to remind them about my illness and outlined again what had happened and who we held responsible at the Trust. Having to relive everything through completion of the NMC forms could easily cause a relapse or worse and neither she nor the rest of my family were willing to risk that.

We had provided the NMC with a huge volume of information explaining my actions, expressing my regrets and including firm denials of false accusations and conclusions reached. Revisiting those three years of bullying, victimisation and uncharacteristic behaviour would be a horrendous and traumatic task for me.

She had come to the conclusion that, like the NHS trust, the NMC were determined to find me at fault; and, this was supported by their decision to refer this to their Conduct and Competence

Committee (CCC) rather than their Health Committee. Were they trying to suggest that I had never been ill?

They seemed to have reached their decisions without any consideration of the information they had had about my health or my submissions providing mitigation and defence. I had decided to seek Voluntary Removal because justice seemed to be a forlorn hope.

She said that hindsight and a healthier perspective had enabled me to realise that the NHS provides a toxic working environment. And, she had real concerns that the various layers of professionally qualified management at my Trust could fail to recognise someone's ill health and that this should lead to such catastrophic circumstances.

Then, she recounted how those same managers never once contacted me or my family and had never shown compassion or humanity. She had even had to force the HR department to phone the hospital to confirm that I was actually there.

What kind of organisation, never repeat never, tries to find out how an employee or his family is coping with a situation like this? Perhaps they did not believe that I was ill but they still had an obligation to find out.

The NMC had asked where I was now employed.

My wife told them that I was happily back at work in a safe and supportive environment well away from the NHS and that was all they needed to know. I was not using my nursing qualification to gain employment and, so, where I was working was my own business.

Furthermore, I had not practiced as a nurse since August 2012 when I was suspended from work pending further investigation and left at home with no support or updates for several weeks ultimately leading to my attempted suicide.

She realised that the NMC would have to ask my NHS Trust about Voluntary Removal and she expressed pessimism about this in view of the vindictive and vexatious approach my Senior Manager took to my case.

But, if the NMC did decide to hold a hearing, I would not be present nor would I be making any further submissions.

My wife then suggested that the saddest thing about this whole situation was that during my 19 years in the NHS, initially as a support worker and then as a nurse, I had worked with some very disturbed patients and I had never once endangered one of those people. She emphasised that, throughout my nursing career, I had shown great dedication and compassion and that that is one huge loss to the NHS

In contrast the NHS had utterly failed me. She

recounted some of the more egregious episodes from this story and suggested that the actions of my managers and colleagues had been devilry at its finest. She added that, when I had become unwell as a result of workplace stress, I was treated as a delinquent employee which only worsened my condition.

All of these issues were something the NMC should reflect upon seriously before handing down its verdict. If I had written this letter on my own, I could not have expressed my feelings better than we did as a family.

A month passed and the anniversary of the referral to the NMC by my NHS Trust drifted by without any further correspondence. Then, one month after the NMC received my wife's letter, one of their lawyers responded. The best description of the process we entered into would be 'good cop, bad cop'.

The lawyer said that Voluntary Removal was not possible because of the seriousness of the charges and that she would oppose an application to the Health Committee because, as far as they were concerned, I was no longer ill. But, they might be prepared to try Consensual Panel Determination, a sort of plea bargain, if I confessed to all my 'crimes'!

The 'good cop' in all of this was the lady with the

Russian name who contacted my wife the next week to arrange a telephone conversation. This proved to be quite interesting because, at the start of the conversation, she sent an e-mail with a modified 'Schedule of Charge' attached.

The original schedule had included two new incidents put forward by my Trust. These were based purely on double hearsay evidence, e.g. 'John told me that Mary said that she heard Charles say............! There was no substantive or corroborative evidence and these charges were dropped.

Also, included originally, was the ward incident where I was hit over the head by a patient using a video controller. If you remember, my Senior Manager had illicitly extracted the documentation from the HR archives. This should, of course, have been destroyed and not been available to the NMC. Again, this charge was dropped.

What was left was the expenses claim error for £5, the altered MRI appointment letter, a charge about swearing on a course and the famous 'fireman's lift/knee squeezing' allegation.

And, whilst I was reasonably comfortable with 'copping a plea' to the first three items, there was no way I was going to plead guilty to a trumped-up bullying and harassment charge where it was one person's word against mine and where the only independent evidence was in my favour.

So, finally we decided to go onto the attack and produce a fully argued case against the new Schedule of Charges and, just in case, against the three charges which had been dropped.

The family machine went back into action even though everyone was feeling fairly war-weary. But, we sensed that the NMC thought that the Trust's case was very weak and that they were quite desperate to settle things 'out of court'.

Well, they dropped the £5 expenses claim charge but the trust just would not let them drop the bullying and harassment charge. It was indicated that, if I accepted the three remaining charges, the NMC would 'let me off' with a six-month suspension.

There was no way that I was going to accept the bullying and harassment charge but also there was no way that I would be ready to return to nursing in six months. So, we decided to fight on and face a hearing of the NMC CCC who would consider my fitness to practise.

The hearing was scheduled for 4 days in the first week in July 2014. Difficult to believe that it would take so long – we thought. How wrong we were!

From previous experience with my Disciplinary Hearing at the Trust, we realised that someone needed to be at this hearing. However, my wife has a full-time, demanding job; so, my in-laws stepped in and said that they would represent me.

So, at this point, I shall hand over the narration of this story to them. And, to make myself useful, I shall head for the kitchen and make everyone a nice cuppa!

<u>The NMC – Hearing 1</u>

Normally, only one representative could speak at an NMC panel hearing. However, we were able to find a precedent for having two representatives provided only one dealt with each witness. It was not possible to 'tag team' against a witness.

Having prepared very carefully for the hearing, we decided that James should accept the charge of swearing during a training course. However, we were going to challenge why the charge had been pursued. It had never been the subject of a disciplinary hearing at the Trust; and, readers of the first book will know that there were good, training reasons for the use of swearing. But, of course, the charge was literally true and he had disobeyed his Senior Manager.

We also decided that he should accept the charge of altering his wife's MRI appointment letter, but that we would use something called the 'Ghosh defence'. This accepts that the ordinary man-in-the-street would regard what James did as dishonest; but, it asks the panel to accept that he was under such stress and so deep in Reactive Depression that he did not understand what he was doing.

The Trust was fielding three witnesses for the bullying and harassment charge which we were not going to accept. So clearly, they had to be cross-examined within an inch of their lives.

We also decided that we would 'push the boundaries' with the Chairman in an attempt to get as much information, about the treatment James had received from the Trust, over to the panel during the course of the proceedings. We hoped that our lay status might persuade the Chairman to be kind to us.

The CCC panel consists of three people, one of whom is Chairman. Generally, two are lay people and one is a qualified nurse in the discipline of the 'Registrant' (James). There is a 'Case Presenter' who is like a prosecuting counsel and a junior barrister usually fills this role. There is also 'Legal Assessor', a more senior barrister, who can advise the panel, the Case Presenter and the registrant or representatives of the registrant.

On day one, a Monday, it became apparent why 4 days had been allocated for the hearing. By the end of the day, a load of procedural stuff had been completed including asking them to allow us to be joint representatives, challenging the relevance of the first charge, making pleas, initial submissions and points-of-order.

We spent more time out of the room than in it whilst the panel made various minor decisions. The problem is that the right of appeal from these hearings is to the High Court; so, the standard of conduct and record has to meet the court's requirements.

Although we had accepted the first charge on James' behalf and no evidence was necessary, the panel had faced a dilemma on the first day. A witness had been brought 300 miles to give evidence that James had sworn on the training course and so they decided that they had to hear him.

What a terrible waste of money and time when we had already indicated to the NMC that James would not contest the charge. But, we saw the 'open door' and took the opportunity to talk about the bullying emails from James' Senior Manager around that time. We also suggested, yet again, that this charge should never have been brought to the NMC in the first place as it had never been the subject of a disciplinary hearing at the Trust.

On day 2, the panel moved on to the second charge, the MRI appointment letter. We made a submission about the 'Ghosh defence' and the Legal Assessor was given time to make an advisory submission about this. During his preparation for this, he made the first of a number of sympathetic visits to the preparation room allocated to us.

He indicated that he was going to say that the argument we had made was a perfectly valid one. However, his advice to the panel was that they had to make a choice between James being ill at the time or that his 'moral compass' had deviated because he was under stress.

Sadly though, the panel would have to make this choice without any independent expert medical evidence because only the Health Committee could call for that. We had provided plenty of medical evidence but it was not 'independent' – Catch 22 or what!

The Legal Assessor gave his advice and by this time there was a break for lunch. The afternoon was to be devoted to the bullying and harassment allegation; and, this was where the planned timetable really went to pieces. Admittedly, this was partly down to our having prepared some very close and detailed cross-examination; but, it also became clear that the panel thought some of the evidence was dodgy which prompted them to ask some difficult questions.

The first witness was James' immediate line manager who had made such a mess of the initial stages of the allegations made by his female colleague. He did not do any better under cross-examination and only served to confirm a number of the points our side had already made.

The second witness was James' female colleague. Her credibility started to collapse when, under cross-examination, she started to try and justify herself by adding to her original testimony and exaggerating. Then one of the panel members asked a couple of questions that finished her off.

First, he asked whether she had been lifted in full view of course participant. She said that they had been crowded round her and had fallen silent with shock. He then pointed out that there were five people who had participated in the course, who were randomly chosen by the Trust investigator and who had said that they had not seen the incident. She countered glibly that they must have been James' personal friends; that did not go down well with the panel.

Then, he asked her, if she was so mortified, why did she not leave the room and seek out a manager immediately. She had no answer to that. It was clear to us that the body language of the panel had changed dramatically during this testimony.

And then, they moved on to the evidence of James' Senior Manager; this was read out and the Case Presenter asked questions. By then, it was time to pack up for the night.

On day three, our daughter was also at the NMC because she was giving evidence on her husband's behalf; and, it was clear that if we had failed to get any messages over during our handling of the case, she was going to remedy that!

James' Senior Manager returned to the room and we carried out a very rigorous cross-examination. Four times the answers were 'economical with the truth' and four times the truth was nailed down. Then, the same panel member asked the witness why in the written evidence there was one paragraph which was identical to one in James' line manager's statement all but for one word.

The answer was that the evidence had been given over the telephone and that it must have been the investigators who used the same paragraph. The panel were aghast that the investigation had been carried out over the phone and were very doubtful of the answer.

They moved on to our daughter's evidence – she was our only witness. She gave a very moving, quite emotional account of the entire episode seen from her viewpoint. The panel were kind and understanding and you could tell from their questions that they were sympathetic.

Some procedural issues padded out the third day and then they packed up for the night. Final submissions were to be made the following morning but it was becoming clear that there would not be time to finish everything on the fourth day unless the pace was stepped up.

In fact, the final submissions were finished by mid-morning and the panel went 'in camera' to consider whether the allegations were 'proven'. Just after midday everyone was called into the room again.

The Chairman announced that, whilst they had made their decision on whether the allegations were proven, there was not time to write-up the decision before they had to finish. Now, as they were being paid for working the whole of the Thursday and as we had committed ourselves to be there for the whole day that seemed very strange and extremely inconsiderate. Why couldn't they at least let us know which charges were proven before they went off for their extended weekend?

Yes, the panel still had to decide whether the actions in the proven allegations were misconduct, they had to decide whether James was 'impaired', they had to hear mitigating evidence and then they had to decide whether there was a sanction. For each stage, there could be a submission from the Case Presenter and from us. To say the least, an NMC hearing is a slow and ponderous ritual. But, the most important thing for us was to know whether the charges were proven and, at that point, they knew the answer.

Nevertheless, the Chairman confirmed that they would do no more on that day and he asked us to find three more days to put in the diary. We were eventually able to find three days in October to suit everyone.

To cut a long story short, everyone turned up in the October. Allegation 1 about the training course swearing was proven because it had been admitted. Allegation 2 about the altered appointment letter was also found proven but there was no decision at that point on the 'Ghosh defence'. However, Allegation 3, the bullying and harassment allegation was not found proven and so the family decision to attend the hearing and fight the good fight was vindicated.

We made many of the same arguments that we had made in July about the flaws in Allegation 1 and the 'Ghosh defence' for Allegation 2. We won on Allegation 1 and it was not judged misconduct. Almost inevitably, Allegation 2 was found to be misconduct.

James was found to be 'impaired'. The panel had already heard the medical evidence once during the 'Ghosh defence'; but, they were forced to endure it again at the mitigation stage. There were also four supporting letters from former female colleagues of James.

James received the sanction of 1 year's suspension from the register. And, he would be expected to write a new reflective statement at the end of the year if he wanted to escape the suspension. Did they really expect him to want to keep reminding himself of what had happened to him?

It will not have escaped the readers' attention that he was offered a lesser sanction during the plea-bargaining stage. However, it was important to get rid of that bullying and harassment allegation; and, James was in no fit state to make a decision about a return to nursing for at least another year.

<u>Trust Chairman No. 1</u>

Having stabilised the situation for a year, and finding that we were 'geared up' and had a little time on our hands, we turned our attention to James' former employers and to how they could be persuaded to recognise the failings of James' former Senior Manager. By then, we had very little regard for the Trust management team; so, we decided to write to the Chairman of the Trust and to try to gain the attention of the Board.

The opening volley was the allegation of a massive waste of Trust money and resource, abuse of Trust processes and clear and vindictive group bullying and harassment of a vulnerable employee. Over three, very full A4 pages, these allegations were expanded and the Senior Manager and her senior colleagues who aided her were identified.

Then, for each of his Senior Manager, his line manager and his 'training buddy' a comprehensive list of allegations was attached with cross-references to the relevant clauses of the NMC Code and/or Trust rules and procedures and/or relevant laws. For his Senior Manager, the list of allegations stretched to three A4 pages, closely typed. And, we pointed out that the allegation which had led to James being fired had been disproved at the NMC hearing.

Just over a month later, we had a letter back

from the Chairman. He completely ignored the allegations we had made. Instead he focused on the Trust's disciplinary processes.

In short, he said that James had exhausted all the internal rights to appeal. He added that James could have gone to the Employment Tribunal. He denied that the NMC panel decision on the bullying allegation was relevant although he could not stop himself from making a triumphant oblique referral to James' suspension.

So, we wrote again providing even more detail and offering to send him evidence for each of the allegations we had made. He obviously had not read any of the materials we had sent because he said that we had raised nothing new and so there was nothing further or new for him to add!

At the same time as we first wrote to the Chairman we submitted another Data Protection Act Subject Access Request (SAR). We thought it could be useful to find out what had been going on behind the scenes since our last request.

The SAR was very timely as we picked up a copy of the internal enquiry which had been carried out for the Chairman to enable him to answer the letter. In the best traditions of independence and impartiality, the internal enquiry had been carried out by James' Senior Manager and, what a surprise, the finding was innocent on all counts!

This same NHS Trust was already getting a bad name for itself for its insensitivity to vulnerable patients and its bullying of the families of patients who had died in unfortunate circumstances. It was becoming increasingly obvious that the Trust was rotten throughout.

<u>Don't the same rules apply?</u>

Clearly, if the Trust would do nothing about flagrant breaches of its rules, policies and procedures, the answer had to be to complain to the regulatory body. So, we submitted referrals to the NMC for James' Senior Manager and for his 'training buddy'.

The referral for his Senior Manager listed over 100 allegations and we provided about 200 pages of supporting evidence. Many of the allegations related to issues which were regarded as gross misconduct in the Trust rules, policies and procedures. All of the allegations were breaches of the NMC Code.

It took the NMC nine months to consider the referral even though it was put through a new experimental 'fast-track' procedure involving Case Examiners. These ineffectual Case Examiners did not examine or investigate. They just asked the Trust for their opinion and enquired whether the Trust was prepared to take Disciplinary action.

Needless to say, the Trust would not take any disciplinary action and gave their full support to their manager. So, the Case Examiners concluded that James' Senior Manager would not be found 'impaired' by a Conduct and Competence Committee. This only confirmed our view that the

NMC exists simply to do the bidding of NHS Trusts even though it is paid for by the long-suffering nurses.

The NMC also mentioned that the Case Examiners' decision had been influenced by four letters of support from colleagues. These were from: a newly appointed clinical colleague who had been given responsibility for training oversight; the Senior Manager's boss who had rejected James' grievance; a senior direct report; and James' former line manager.

Regardless of the quality of this mitigating evidence which was clearly not very good, the NMC rules quite clearly state that such evidence should not be considered until a decision on impairment is made by a fitness to practise committee. This rule was blatantly disregarded.

Nevertheless, our appeal directly to the Registrar (the CEO) of the NMC was rejected. It seems that there is one rule at the NMC for senior managers of Trusts and another for the ordinary nurses.

The referral for the 'training buddy' was a more tortuous affair. Firstly, it turned out that, a few months after James' departure, a 'mutually agreed' departure from the Trust was arranged for this former colleague who had done so much damage and, of course, who had some damning information about the activities of the Senior

Manager and the team. The Trust said that it did not have a forwarding address that the NMC could use!

We did a simple Google search and found a way for them to contact him. Eventually, the NMC did manage to do this and he immediately submitted a request for Voluntary Removal from the NMC register. Now, when James asked for this, he was told that it was not possible whilst he was being investigated.

This did not stop the NMC from removing his 'training buddy' from the register in breach of its own rules. When we protested, the NMC said that there was nothing that could be done now that he was off the register. How convenient!!

The rule bending, if not breaking, with these two NMC referrals was breathtaking. A conspiracy theorist might well be going into overdrive at this point but it all seemed to be part of a pattern!!

Whilst all this was going on, we learned that James' former line manager had also recently ceased full-time work for the Trust. So, the group of people that James' Senior Manager had so contemptuously described as an 'inherited team' in 2010 was finally disposed of!

<u>The NMC – Hearing 2</u>

The NMC scheduled a review hearing for James on 7th October 2015. He was not ready to return to nursing and, in fact, was almost certain that he did not want to return ever! But, the family prevailed on him to give himself another year.

James wrote a very polite letter to the NMC stating that he was still recovering from his illness and would be grateful if they could extend his suspension for a further year. No questions were asked by the NMC Case Manager and no further information was requested. Well they had all the information about his illness in their internal filing system didn't they?

We had indicated that no-one would be present on this occasion and so we relaxed to await a positive answer. Like a bolt out of the blue, we received the findings of the NMC panel including a transcript of the proceedings.

The Case Manager had not provided the medical evidence from the first hearing to the second panel. On reading the papers for the hearing, the Case Presenter, a reasonably senior barrister, had not queried the reference to illness and the lack of evidence. When asked by the panel Chairman what evidence there was, the Case Presenter denied there was any without thinking about checking his facts.

The panel found James to be lying about his illness and therefore even more 'impaired' than he had been before. They gave him three months to appear before another hearing.

At this point, several members of the family went ballistic. James' wife spoke to the NMC Head of Legal Affairs who removed the website link to the hearing information and agreed that a new hearing would be scheduled for December. James' father-in-law complained to the NMC Chief Executive and secured a personal letter of apology to James.

The excuse given was that the NMC scanning department failed to scan in the medical evidence after the first hearing. But, the evidence must have been on the NMC servers because it was part of the defence evidence bundle. And, anyway, surely someone should have had the intelligence to spot that something was missing before the hearing. A quick e-mail to the family could have solved the problem.

It is perhaps typical of senior people in an organisation like the NMC to blame their juniors rather than accepting that they are at fault.

<u>The NMC – Hearing 3</u>

The family tag team of James' in-laws embarked on preparations for the December hearing. This was not to be a meek and mild affair. We decided that the NMC panel would get all the gory details and that what had once been intended to be a simple request for a one year extension would be another opportunity to make our feelings known about the way James had been treated both by the Trust and the NMC.

A particular target was the NMC Case Presenter from the October hearing. As an experienced barrister, he should have asked the Case Manager to look for the medical evidence. Furthermore, when the panel Chairman asked him where the evidence was, he should have asked for a short recess rather than making an adamant denial that there was any evidence.

But, it became quite clear that the panel were in no mood to upset us or argue with us. They had been briefed that the NMC had made a significant error and we got the result that James had asked for.

We did decide though that it would not be enough to just send a letter before the next review hearing; we would have to be present.

<u>Trust Chairman No. 2</u>

The Trust Chairman who had answered our previous correspondence retired in mid-2015 and a long-established member of the Board took his place. We were actually able to speak to him at the January 2016 Trust Board meeting which was attended by many disaffected families – mainly the families of mental health patients who had died whilst in the care of the Trust.

He seemed a reasonable man and so we decided to renew our attempts to get something done about James' Senior Manager. To keep it simple, we sent an e-mail attaching the correspondence with his predecessor, pointed out that the only investigation into the Senior Manager's actions had been carried out by that Senior Manager, and offered to send all our evidence if he would institute a proper investigation.

Much to our surprise, the same day, we received an e-mail reply saying that he wanted to familiarise himself with the issues and then to arrange a meeting so that he could understand our position.

But, nearly two weeks later, we found that nothing had changed. The Trust executive who wrote the letters for the previous Chairman had clearly got a grip on the situation. The new

Chairman's response differed little from his predecessors. He claimed that 'relevant internal investigations' had taken place and he was 'assured that the processes followed throughout were proper and robust'.

It is perhaps telling that only a couple of months later this Chairman was asked to resign and he was replaced with an Interim Chairman appointed by NHS Improvements. We were not quite aware at the time that he was only the first domino to fall.

<u>Trust Chairman No. 3</u>

The new Interim Chairman started well by holding meetings with disaffected families and with some of the Trust governors who had been calling for action. He even suggested that 'the families' should put their views about the people at the top of the Trust to the Board of Governors.

As the family of a former staff member who was bullied into Reactive Depression and a serious suicide attempt, we were slightly different to the families of patients who had died because of poor care. But, we thought 'in for a penny, in for a pound'.

In July 2016, we sent an e-mail to all the Trust governors, to the Interim Chairman and to several interested politicians and ministers. We took a more general focus by criticising staff management and welfare and asking how a Trust could provide appropriate training if the training policy was being controlled by someone like James' Senior Manager. We provided everyone with a detailed 'Exposé of Bullying' at the Trust.

The new Interim Chairman immediately e-mailed back and said that he was going to appoint an independent investigator. He asked for any evidence we had to be provided to him by the month end. The following day we sent a paper copy of our 'Exposé', our evidence bundle for the NMC referral and some NMC hearing transcripts which showed that the Senior Manager had tried to mislead the panel.

We did not get too excited but we felt that someone seemed to be taking us seriously. However, come September we were disappointed yet again. We were told that the investigator had reported that there was 'no evidence to support the allegations made'.

When we examined the report, we found that it was written by a 'gun for hire' character assassin who had carried out other investigations against NHS whistle-blowers and bereft families. The entry of the investigator's name into Google produced a lot of not very complementary comments.

The supposedly 'independent investigator' had looked at just four pages of our 200 pages of evidence, had picked holes in two of them and had pronounced us and, by association, our son-in-law as unreliable.

We were furious but this was tempered slightly when the Interim Chairman, the Chief Executive, the Chief Operating Officer and the boss of James' Senior Manager all disappeared from the Trust. The dominos were continuing to fall.

A new and much more business-like Interim Chairman was appointed and a new Interim CEO was put in place. They started to talk seriously to the families and to current and past governors to find some solutions to the Trust's problems. Also, we had developed useful links with governors, families and other activists which enabled us to continue lobbying for change.

The NMC – Hearing 4

We approached the October 2016 review hearing with some concern because we were not quite sure how to end 'the never-ending story'. James had finally decided that he never wished to return to nursing. All he wanted to do was to get off the NMC register.

However, the NMC rules do not allow a registrant to get Voluntary Removal from the Register until any sanction they have been given is finished. And, the hearing was being held before the suspension would end.

So, the logical thing seemed to be to ask the review panel to recommend removal from the register as soon as the suspension ended. This we decided to do but, when we discussed this with the Legal Assessor, he said that, under the 'accepted interpretation' of the NMC rules, the panel could not agree to this.

The NMC Case Presenter had been instructed to call for a further year's suspension but he could not see why that was a good idea and nor could we. We asked whether this meant that the nightmare could go on forever.

Luckily, the Chairman of the panel was very sympathetic and we took the risk of recounting James' entire story all over again. And, we pointed out that he had made a very contrite reflective statement at his first hearing; he also had written a very sensible letter to this panel explaining why he could not return to nursing.

It became clear that the entire panel was very sympathetic. Under the guidance of the Legal Assessor, they decided that the original reflective statement combined with his letter showed that James was no longer impaired.

They reached this conclusion right at the end of the day and had not had time to write the decision up. Nonetheless, they told us and sent the written decision later. Quite a contrast to the behaviour of the panel at the first hearing!

So, because the NMC rules are so completely stupid, James was freed to return to nursing even though he did not wish to. Surely the time has come for the rules to change so that people can drop out of nursing on a permanent basis without jumping through artificially constructed hoops?

Very shortly after the hearing we learned that some of the rules are to change through a new Act of Parliament – that is once parliament has time to deal with something other than Brexit. Mind you, it does seem daft that these changes cannot be made by secondary legislation approved by the Secretary of State for Health.

I suppose that some politicians in the past wanted to get an ineffective NMC under control;

but, the NMC is now clumsy, slow and ineffective because it is in a legalistic straitjacket. Instead of making small procedural changes on an evolutionary basis like most private sector organisations, it has to make do and mend with half-hearted experiments like the Case Examiners whilst waiting for parliamentary time. The danger now is that more time and money will be spent on lobbying and influencing parliament than on running the organisation.

One of the changes proposed we regard as a major victory for our lobbying efforts. The Conduct and Competence Committee and the Health Committee are eventually to merge. The combined Fitness to Practise Committee will be able to call for Legal and Medical advice for any hearing.

If that had been possible for James things could have been very different. How many more people are going to have to suffer as he did before these changes are implemented?

<u>A Good Bedside Manner</u>

One of the Trust governors worked hard on our behalf and secured a meeting with the Interim CEO. He said that, as a registered nurse, she had a much better attitude than her predecessor who was a professional administrator. He was sure that she would at least give us a fair hearing; he felt she had 'a good bedside manner'!

We were booked in for a one hour appointment on a day in early January 2017 – six years out from the start of James' troubles. 'We' being James' wife and us, her parents. After all this time and bearing in mind our past experiences, we did not expect to achieve much but we received a warm welcome.

In fact, she gave us a generous 2 hours of her time and listened intently whilst the three of us gave our sides to the story. She had read the 'independent investigation report' but, after listening to us, she admitted that it was neither independent, nor an investigation nor a proper report!

She explained, though, that, as an interim CEO, she faced a number of problems. Firstly, in addition to the departure of some very senior executives, the Trust had lost a lot of people through poor morale. And, the reputation of the Trust was such that recruitment was extremely difficult.

Also, a high-level strategic review was underway and it was possible that the shape of the Trust could change quite dramatically. So, she could not be sure how many people and of what seniority would be required.

Alongside that, she had looked at James' Senior Manager's file and the appraisal reports were good, there were no other complaints of bullying (after what had happened to James that was no surprise) and no disciplinary issues. So, the only justification for a full disciplinary investigation was our complaint which had been turned down by three Chairmen.

Before the appointment, we had discussed, as a family, what we wanted from our meeting. We had already faced up to the possibility that, 6 years out, it would be difficult to secure another, full-scale investigation let alone disciplinary action. We might want this domino to fall but we had to be realistic.

So, we decided that the least we wanted was for the CEO to meet with the Senior Manager and discuss what had happened. Then we believed a suitable file note should be placed in the personnel file. The Senior Manager should be left in no doubt that a repeat performance could lead to dismissal.

We explained what our wishes were and the CEO agreed that that is what she would do. A couple of weeks later she confirmed that the meeting had been held and the note had been inserted in the file.

Perhaps this was not a full measure of justice but, with all the other uncertainty that was being felt within the Trust, it was probably more of a punishment than it might seem on the surface.

We also knew, from the two DPA SARs that we completed, that the Senior Manager had read 'The NHS, the Uniform & the Blackberry®'. The comments about it that we saw were angry ones but self-recognition can sometimes lead to a change in someone's behaviour; and, anyway, that anger would have been a punishment in itself.

A little unexpected bonus, a few weeks after our meeting with the CEO, was that all the Trust Non-Executive Directors were asked to resign and the person suspected of writing the negative letters for the Chairmen left the Trust – yet another set of dominos had fallen!

The Very Last Word
(from the Author)

As I started writing this last chapter, a very brave Prince Harry carried out a podcast interview with Bryony Gordon for the Sunday Telegraph. His stated aim was to break the stigma surrounding mental health. How appropriate and look at the fascinating parallels it provided.

Prince Harry lost his mother in tragic circumstances at the age of 12 and he admits to shutting down his emotions for nearly 20 years. James lost his father in tragic circumstances at the age of 13 and he now knows that he shut down his emotions for about 23 years.

Prince Harry was experiencing extraordinary pressures living in the public eye and came very close to a complete breakdown on numerous occasions. James was facing extraordinary pressures in his home and work life and actually experienced a complete breakdown.

Prince Harry talked of two years of total chaos in his life and of being on the verge of punching someone. James experienced about two years of total chaos in his life and certainly let grief, worry, anger and frustration make him behave in an erratic and uncharacteristic manner.

Prince Harry credits his close family with persuading him to find counselling. Having been able to talk honestly about his feelings he says that he now feels able to put "blood sweat and tears" into making a difference for others.

James family helped him to get counselling in the absence of timely support from his NHS Trust. Having been able to talk honestly about his feelings, he is now able to contribute to the community and to hold down a job again.

I have seen the excellent reference James received from his university tutor after he graduated in 2004; I have seen the superb reference given by his Modern Matron in 2006 when he applied for the training job; I have seen the encouraging feedback from trainees attending the courses he has run; and, I read the feature article on him in the Trust house magazine in 2010.

Two organisations he volunteered with, when he was recovering in 2013, immediately spotted his potential. Then, his new employers quickly placed him in their management training programme and, so far, have given him two performance awards.

So, what went so wrong in 2011 and 2012? James was the same person and he had the same potential. But, the atmosphere at work became poisonous and the law of the jungle prevailed. That combined with major stressors in his private life drove him into severe Reactive Depression.

What is really disturbing is that, faced with such an obvious case of Workplace Stress and Reactive Depression, both management and nursing professionals at James' NHS Trust should have been so completely blind to the needs and feelings of a troubled colleague. In the information gathered from the two DPA SAR outputs, there was not even a mention of the possibility of stress, depression or another mental health issue being behind his behaviours. And, his Senior Manager callously said "Well, I am not a mental health nurse".

James' NHS Trust let him down as an employee and as a patient. It is a mental health Trust and yet, for staff suffering from mental health problems, it is a toxic environment. When James' problems did come to the surface, they ignored them. And, when he required counselling, they could not provide it when he needed it.

What a gross waste of talent! At a time when the NHS is in dire need of every qualified nurse and doctor it has, one of them gets treated in this way.

We started out with the simple objective of highlighting the bullying culture in the NHS by describing the actions of one senior management bully. But, the failings in the NHS go deeper than that and, sadly, the bullying culture seems to be endemic.

That is where the overall title of this combined book, 'The Busy, the Bossy and the Bully', is so apposite. Never have the nurses, doctors and their support staff in the NHS been so **Busy**. But, their managers believe that the only way to get the work done is to be **Bossy** which makes many of them, like James' senior manager, act like bullies.

However, it is the NHS system which is the real **Bully**. James' story highlights how the first norm is to punish not to try to understand. Orders are given without explanation as though the recipients have no experience and nothing to contribute. Deviation from the 'true path' is frowned upon and internal whistleblowers are treated as treasonous rather than being listened too.

Formal disciplinary action is a first resort and, from a Freedom of Information request (FOI) completed during the writing of this book, NHS Trusts waste millions of pounds on their internal disciplinary procedures.

Governance in NHS Trusts is too big an issue for this book but until Governors of NHS Trusts have some governance powers and until Non-Executive Directors outnumber the Executive Directors, there will continue to be inadequate governance in the NHS and a cavalier approach to the use of public money.

Once NHS Trusts have found someone guilty,

instead of coaching and retraining they use the cosh of a referral to the NMC. NHS Trusts refer far too many people to the NMC and, because it is overwhelmed it does things badly. In truth, the regulatory system for nurses and midwives is just not fit for purpose. The NMC has become part of the **Bully**.

Some changes in NMC procedures have been put out for general consultation but, even if they are implemented, the whole system will continue to be a sledgehammer to crack a nut. One of the biggest problems is that the NMC employs too many lawyers and paralegals and not enough people who understand nursing, business management or human nature. Until that changes, there will continue to be a problem.

A large proportion of the people that go through the NMC process, and that is some four to five thousand a year, never return to nursing. This is not because they have committed some heinous crime but because the very regulatory process is so time consuming and unpleasant.

And yet, the NMC regulatory system is at its weakest for the worst offences. The maximum sanction is to strike someone off the register for five years, which is very little problem for a nurse who commits manslaughter and might have to spend that time in prison anyway.

However, if you are a nurse referred to the NMC

by a Trust for a much lesser offence, as James was, you can spend at least three years in a state of suspended animation before you can return to the vocation of your choice.

James' only offence that received a sanction was that he altered his wife's MRI appointment letter so that he could get leave to accompany her to that appointment; this was because she suffered from claustrophobia. He was suspended from his job at the NHS Trust in August 2012. Even if he had chosen to return to nursing after a 1 year NMC suspension, he could not have returned to work until the end of 2015. Does that make any sense when the NHS is so short of nurses?

For this one case, the cost of employment of the NMC staff involved over four years, the fees for the external lawyers and the costs for the conduct of 11 days of NMC committees, during that time, must have amounted to at least £200,000 on a fully allocated cost basis.

Is it any wonder that the government had to inject £20 million of taxpayers' money in 2012 just to keep the NMC afloat prior to the huge hike in nurses' annual registration fees?

The NMC has an obligation to its paying registrants to recognise illness and disability and to modify its processes and procedures to help them through the trauma of referral and investigation. It even has a Board director dedicated to mental illness. But, until it does change its rules, it just cannot handle someone who slips in and out of Reactive Depression.

Isn't it time, too, for the NMC to be forced to become accountable to its paying registrants rather than acting as a lapdog of the NHS trusts? It does seem as though registrants at the NMC are, effectively, buying the bullets for their own firing squad! From the start, the NMC treated James like someone who was going to be punished not as a registrant who had had unproven allegations made against him.

Surely, NHS Trusts should have to make a substantial, partially refundable payment to the NMC when they make a referral. Then, if the referral fails, there would be no refund; but, even successful referrals would largely be paid for by the organisations which are demanding regulatory support.

In the main, the NMC acts as a very expensive 'Director of Nurse Prosecution' for NHS Trusts not as an effective investigative and regulatory body. The NMC has a responsibility to its paying registrants to check every piece of evidence submitted by NHS Trusts as they would for any other complainant; and, they should use properly qualified investigators not paralegals from local firms of solicitors

If anything, the current approach to Trust referrals stifles good staff management in those Trusts by providing a convenient punitive alternative to the hard graft of encouragement, regular appraisal, continuous review, personal coaching and quality training.

Currently, when it comes to fitness to practise referrals, the NMC is a bit like a road scarifier. Very slowly, ponderously and deliberately it shreds nurses' careers. It tears their lives to pieces and leaves it to someone else, if they are lucky, to put them back together again.

I do hope that the publication of this book might help to promote change and ensure better treatment and a better form of justice for nurses and midwives in the future. But, as the title of this half of the book says – Justice is a Long Road!

www.ingramcontent.com/pod-product-compliance
Lightning Source LLC
Chambersburg PA
CBHW071446180526
45170CB00001B/480